THE PRACTICE OF ROYAL ICING

THE PRACTICE OF ROYAL ICING

By

AUDREY HOLDING

Illustrated by

JOHN HOLDING

ELSEVIER APPLIED SCIENCE
LONDON and NEW YORK

ELSEVIER APPLIED SCIENCE PUBLISHERS LTD
Crown House, Linton Road, Barking, Essex IG11 8JU, England

Sole Distributor in the USA and Canada
ELSEVIER SCIENCE PUBLISHING CO., INC.
52 Vanderbilt Avenue, New York, NY 10017, USA

WITH 113 ILLUSTRATIONS

British Library Cataloguing in Publication Data

Holding, Audrey
The practice of royal icing.
1. Icings, Cake
I. Title
641.8′653 TX771

Library of Congress Cataloging-in-Publication Data

Holding, Audrey.
The practice of royal icing.

Bibliography: p.
Includes index.
1. Cake decorating. 2. Icings, Cake. I. Title.
TX771.H67 1987 641.8′653 87-5315

ISBN 1-85166-086-0

Printed in Great Britain at the University Press, Cambridge

Foreword

When in 1980 I wrote the foreword to *The Art of Royal Icing* I wished my friend Audrey Holding every success for her book; I was convinced that its appearance would fill a real gap in the literature then available by providing a comprehensive but simple and practicable guide to the techniques used in royal icing. Now, six years later, with *The Art of Royal Icing* sold widely throughout the world, I know that Audrey's second book on this subject is assured of success. In it she has developed many of the ideas presented in her first book, and she takes the cake-decorator through the stages of intricate techniques which would normally be regarded as beyond the competence of the amateur. With Audrey's guidance (and lots of patient practice!) we can all become experts.

JUNE ELWOOD
M.Inst.B.B., M.C.F.A.(C.G)
Cake Artistry Studio,
Swinton,
Manchester, UK

Preface

My first book, *The Art of Royal Icing** set out to explain the craft to those who had no previous knowledge of the subject. In this book, I hope to go one stage further and expand the ideas and techniques formulated in *The Art of Royal Icing*, especially in relation to runout figure piping.

In order to avoid repetition, some basics have been omitted, enabling other areas to be dealt with in greater detail. Where it is felt that some reference must be made to a subject covered in the previous book, the details are described as briefly as possible.

In response to requests from readers of my first book, I have included as many drawings as possible that can be traced directly from the book. Readers will also find that most figure drawings are shown in several sizes, thus making them suitable for cakes of different dimensions.

Once again the accent is on edibility; no artificial decorations are included. Piped borders are shown in *The Art of Royal Icing*, but in this book all borders are made using the runout method.

It is not my intention to replace practical lessons; developing your own style is achieved by learning as much as you can from as many sources. I sincerely hope you will derive much pleasure from this book and will also pass on to others the art of royal icing.

AUDREY HOLDING

* *The Art of Royal Icing* by Audrey Holding, Elsevier Applied Science Publishers Ltd, London, 1980, ISBN 0-85334-860-X.

Acknowledgements

I would like to take this opportunity to thank all who have helped to make this book possible; fellow students from College days who are still close friends and continue to share their ideas; new friends who tackle cake decorating with such enthusiasm and students who constantly come up with techniques that I should have thought of years ago!

Most of all I would like to thank June Elwood and her husband, Dr Willis J. Elwood, who have been my main source of help and encouragement.

The author would also like to thank the following for permission to reproduce illustrations and other material: Marjorie Berry, Barbara Buckley, Doris Chatwin, Daily Express, Simon Elvin Ltd, Sheila Excell, Fine Art Developments p.l.c., Bunty Gibson, Heron Arts, Intercontinental Greetings, Mary Jackson, Roy Kerry, Elaine MacGregor, Florence Parkes, Janis Pryce, Save the Children, Maureen Smith, Peter Stott, Flo. Thompson, Mary Tipton, Andrew Valentine Ltd, Valentines of Dundee Ltd, F. J. Warren Ltd.

AUDREY HOLDING

Contents

List of Illustrations

List of Colour Plates

(Colour plates can be found between pages 120 and 121)

CHAPTER 1

Equipment

Until recently it was almost impossible to purchase specialized equipment for cake decorating. This situation has however changed and most towns now have shops supplying cake decorating requirements.

It is not necessary to spend a lot of money. For example, though I use the most expensive piping tubes, I need only a few and their superior quality means they do not need constant replacement.

If possible, do start off with the right equipment. For instance, do not be tempted to try to get a smooth coating on the cake by using a palette knife. You will need a stainless steel straightedge which will, with practice, give the desired result. Do not be tempted to use a plastic straightedge; it will bend making it impossible to obtain a flat surface. Both straightedges shown in Fig. 1 are suitable, though the one with the flange is superior as this prevents flexing in use.

The equipment required is as follows, and it must be stressed that all utensils used for royal icing must be entirely grease free. Additional items which will be required when applying marzipan or almond paste are listed in the appropriate chapter.

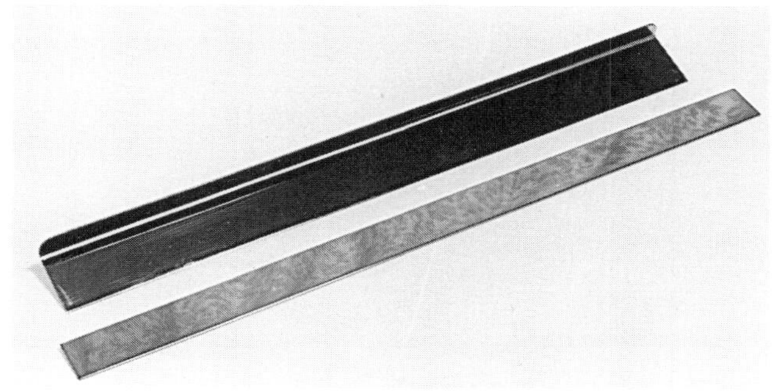

Fig. 1. Straightedge.

Bowl or basin
Large enough to allow icing to be hand-mixed without overflowing. When mixing icing, glazed earthenware or glass (Pyrex) bowls are best as they are more rigid than plastic. Icing can however be stored in polythene bowls (Tupperware) provided they have air-tight lids.

Wooden spatula
This is used to mix the icing. In order to keep it grease-free, do not use it for any other purpose.

Dishcloths
Several will be required.

Sieve
Icing sugar will need to be sieved before use.

Tablespoon
A large spoon will be required when sieving sugar and during the mixing process.

Palette knives
I use two and both have stainless steel blades. A 10 cm (4 in.) knife is used to coat the sides of

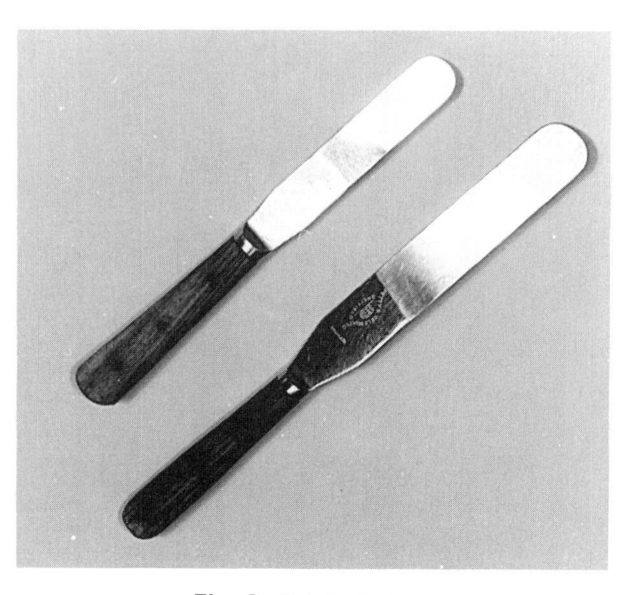

Fig. 2. Palette knives.

cakes; to scrape down the bowl of icing; and to fill piping bags. A larger 15 cm (6 in.) knife is used mainly when icing the top of a cake (Fig. 2).

Tea towel
A damp tea towel will be required to cover the bowl of icing unless it has been placed in an airtight container. As an alternative, a piece of clingfilm may be used.

Straightedge
As mentioned at the beginning of this chapter, it is essential that you have a good straightedge in order to achieve a smooth finish to the surface of the cake. A strip of stainless steel can be bought or made but must not have rough edges, otherwise marks will appear on the surface of the icing.

Side scraper
This is used for obtaining a smooth finish when icing the sides of a cake. Side scrapers are available in plastic or stainless steel, the stainless steel version being superior.

Greaseproof paper
A good quality paper will be required for making piping bags.

Scissors
A small pair of sharp scissors will be required.

Piping tubes
I use a tube which has no seam and is made from nickel silver. Although more expensive than others it is well worth the extra cost. Throughout this book only four sizes of tubes are used. They are Bekenal numbers 0, 1, 2 and 3.

Paint brushes
Fine paint brushes have many uses and include smoothing linework, assisting the flow of icing in runout work and the cleaning of piping tubes. For these purposes an inexpensive brush with nylon bristle (size 0 or 1) is suitable. For very fine painting on the surface of the cake or on runouts, I would suggest a size 000 brush with sable bristle.

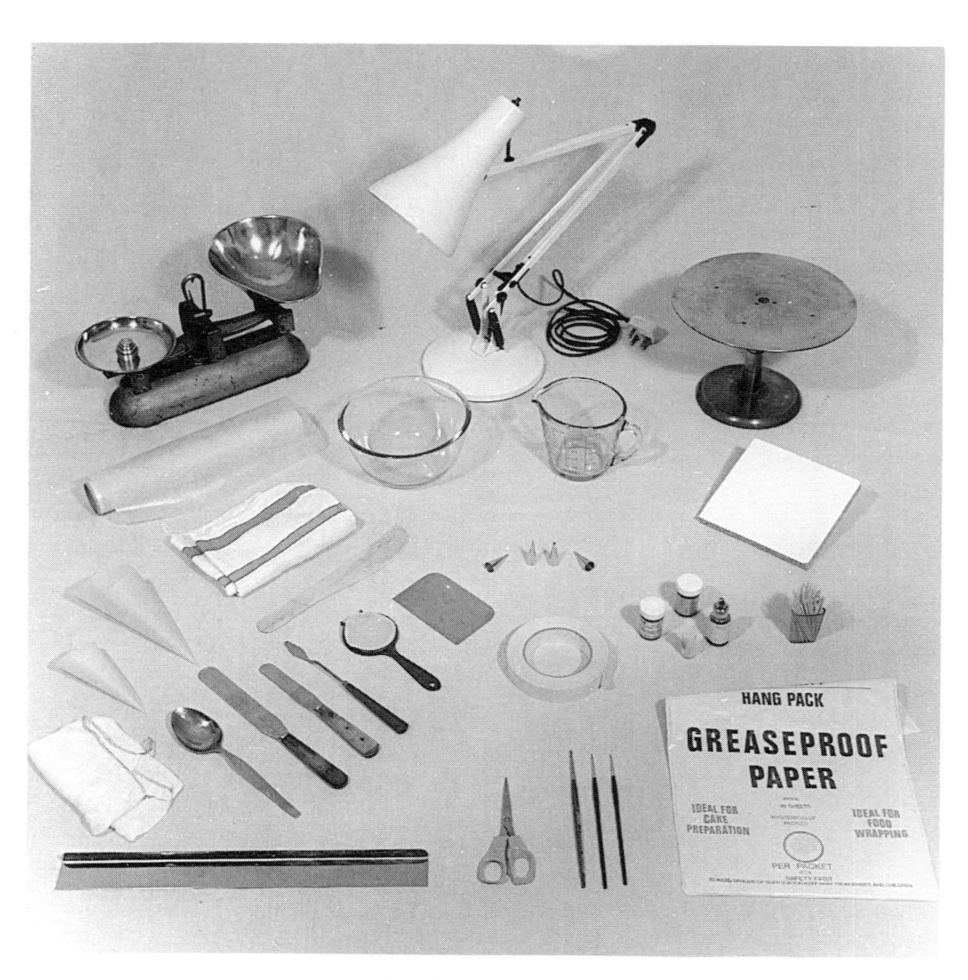

Fig. 3. Equipment.

Turntable
A good turntable is an expensive item but essential if first-class work is to be produced. When choosing a turntable, ensure that it is strong enough to hold a heavy fruit cake and that it rotates smoothly.

Scales and measuring jug
If you intend to use powdered albumen for making your icing, it is essential to have scales and a measuring jug to hand. If you use meringue powder these will not be necessary as the powder will be in ready-to-use sachets and contain a scoop for measuring the water.

Waxed paper
This is required for all runout work. It should be quite strong, but thin enough for the design to be visible through it.

Masking tape
Royal icing can be used for sticking down the waxed paper prior to runout work. Masking tape is also suitable, though more care must be taken on its removal in order to avoid damage to delicate runout pieces. Sellotape is too strong and should not be used.

Colouring
Edible colourings are readily available in powder, paste and liquid form.

Cocktail sticks
These are ideal for transferring colour from container to the icing.

Boards for runouts
Many materials may be used to provide a smooth, flat and non-pliable surface on which to work runouts – for example, an old tile, a piece of glass, or an off-cut of Formica or Perspex.

Artist's palette knife
An ideal tool for lifting small runouts from waxed paper.

Lamp
Direct heat is essential to assist drying and to ensure a pleasing gloss on runout work. An anglepoise lamp is ideal for this purpose.

CHAPTER 2

Making the Cake

Before we go further, attention must be paid to the cake itself. Traditionally, a wedding cake comprises a rich fruit cake covered with almond paste or marzipan and coated with royal icing. It is possible to use royal icing on a light fruit cake, Madeira, or sponge cake, but marzipan or almond paste will still be required prior to icing. A rich fruit cake can, and should be made several weeks in advance.

If you use a sponge cake it will have to be made, decorated and consumed quite quickly as the cake will not keep fresh for long. In my opinion, a rich fruit cake is the ideal base when using royal icing.

Most people who have an interest in cake decorating have already found that they have a talent for baking cakes. Therefore, if you already have your own recipe, use it. For those who would like a recipe, the one I use is shown at the end of this chapter.

Depth or Thickness of Cake

When making a cake, particular attention must be paid to the depth or thickness. If, for instance, you are making a 25 cm (10 in.) diameter cake and intend the finished cake to be 8 cm (3 in.) deep, then the cake alone must be much less than this to allow for the paste. How much paste you apply is for the individual to decide. The amount of mixture per tin suggested in my previous book was less than I now use and the marzipan was much thicker.

The thickness of icing can be discounted; even allowing for several coats, it will not be more than 5 mm ($\frac{1}{4}$ in.).

Individuals vary on how deep they think a cake should be. I do not like tall cakes, and consider a 20 cm (8 in.) diameter cake that is 7 cm ($2\frac{3}{4}$ in.) deep when coated to be correctly proportioned. Obviously, larger cakes should be thicker — about 7·5 cm (3 in.) deep for a 28 cm (11 in.) diameter cake and smaller cakes much less. For example, the 10 cm (4 in.) diameter cake in Fig. 37 is 5 cm (2 in.) deep.

When assessing the depth of wedding cakes it is acceptable for all the cakes to be the same thickness, because when a wedding cake is assembled it is seen as a single cake. I prefer the tiers of a wedding cake to be graduated however; the thickest being at the bottom, the thinnest at the top, and the difference in thickness between neighbouring tiers to be uniform throughout. This gives the cake a balanced look.

Cake Tins

These are available in many different shapes and sizes (Fig. 4), starting at 10 cm (4 in.) in diameter. I prefer a tin not to have a loose base and I avoid buying a square tin that has rounded corners.

Fig. 4. Cake tins.

Never be tempted to over-fill a cake tin, even if you find that the amount of mixture is only slightly too much. Put left over mixture to one side and bake a small cake with it. The cake does not have to be baked immediately and mixture will keep, covered, in a cool place until the following day if necessary. A small cake makes an excellent 'sample' for someone who has not previously tasted your mixture. Small cakes also make ideal gifts for people who live alone.

Lining a Cake Tin

I do not fully-line a cake tin because I find a piece of greaseproof paper, placed in the base of the tin, is adequate. To do this, place the tin on a sheet of greaseproof paper; draw around it with a pencil and cut out, slightly inside the pencil line. This should fit the base of the cake tin perfectly. As I always bake fruit cakes at a low temperature, I find complete lining of the tin to be unnecessary.

Large cakes (28 cm (11 in.) diameter and upwards) may require insulation in order to ensure that the outside does not become burnt before the inside is fully cooked. A simple way of overcoming this problem is to wrap brown paper around the outside of the tin prior to baking. To do this, make a band of brown paper, double thickness, about 5 cm (2 in.) more than the height of the tin. This is placed around the tin and secured with either string or staples (Fig. 5). This band may be removed during the last hour of baking.

Fig. 5. Brown paper around cake tin.

Storing the Cake

Fruit cake should be left in the tin until it is cold. To ensure easy removal, draw a palette knife along the side of the cake tin. Turn out onto a clean surface and remove the greaseproof paper from the bottom of the cake. Wrap the cake in greaseproof paper and seal with tape (Fig. 6). Wrap again in foil and store on a shelf or in a cupboard until required.

Fig. 6. Cake wrapped in greaseproof paper.

Recipe

The following is the recipe I use for making a 20 cm (8 in.) diameter round cake.

Ingredients

225 g (8 oz) plain flour
Half teaspoonful ground cinammon
Half teaspoonful ground mixed spice
Half teaspoonful ground mace
225 g (8 oz) butter
225 g (8 oz) dark soft brown sugar
4 large eggs, beaten
The grated rind of one lemon
225 g (8 oz) currants
225 g (8 oz) sultanas
225 g (8 oz) seedless raisins
100 g (4 oz) glace cherries, halved
100 g (4 oz) mixed chopped peel
50 g (2 oz) nibbed almonds
2–3 tablespoonful brandy

The above ingredients equal one mix in Table 1.

Table 1
Mixture and cooking times for various cake sizes

Approximate number of mixes[a]	*Approximate sizes of resulting cakes*	*Approximate baking time*
½ (4 oz flour, etc.)	10 cm (4 in.) round × 4 cm (1½ in.) thick & 12.5 cm (5 in.) square × 4 cm (1½ in.) thick	1¾ hours 2 hours
½	10 cm (4 in.) square × 4 cm (1½ in.) thick & 12·5 cm (5 in.) round × 4 cm (1½ in.) thick	1¾ hours 1¾ hours
1 (8 oz flour, etc.)	15 cm (6 in.) round × 4·5 cm (1¾ in.) thick & 15 cm (6 in.) square × 4·5 cm (1¾ in.) thick	2¼ hours 2¾ hours
1	17·5 cm (7 in.) round × 5 cm (2 in.) thick & 12·5 cm (5 in.) square × 4 cm (1½ in.) thick	2¾ hours 2 hours
1	17·5 cm (7 in.) square × 5 cm (2 in.) thick & 10 cm (4 in.) square × 4 cm (1½ in.) thick	3 hours 1¾ hours
1	20 cm (8 in.) round × 5·5 cm (2¼ in.) thick	3¼ hours
1½ (12 oz flour, etc.)	20 cm (8 in.) square × 5·5 cm (2¼ in.) thick & 12·5 cm (5 in.) round × 4 cm (1½ in.) thick	3½ hours 1¾ hours
1½	23 cm (9 in.) round × 5·5 cm (2¼ in.) thick & 10 cm (4 in.) square × 4 cm (1½ in.) thick	4 hours 1¾ hours
1½	23 cm (9 in.) square × 5·5 cm (2¼ in.) thick	4½ hours
2 (1 lb flour, etc.)	25 cm (10 in.) round × 6·5 cm (2½ in.) thick & 15 cm (6 in.) round × 4·5 cm (1¾ in.) thick	4½ hours 2¼ hours
2	25 cm (10 in.) square × 6·5 cm (2½ in.) thick	5 hours
2	28 cm (11 in.) round × 6·5 cm (2½ in.) thick	5 hours
2½ (1 lb 4 oz flour, etc.)	28 cm (11 in.) square × 6·5 cm (2½ in.) thick	6 hours
2½	30 cm (12 in.) round × 7 cm (2¾ in.) thick	6 hours
3½ (1 lb 12 oz flour, etc.)	30 cm (12 in.) square × 7 cm (2¾ in.) thick	7 hours

[a] One mix is based on 225 g (8 oz) of plain flour — see previous recipe.

Method

1. Clean fruit and place in a bowl (at this stage the brandy may be added to fruit and left, covered, for several hours or preferably overnight).
2. Grease and line a 20 cm (8 in.) round cake tin.
3. Beat butter and sugar until creamy, then gradually add beaten eggs.
4. Fold in the flour, alternately with the fruit, nuts and lemon rind.
5. Finally stir in the brandy unless this has been previously added to fruit.
6. Bake at 120°C (250°F) or gas mark ½ for approximately three hours. The baked cake will be approximately 5.5 cm (2¼ in.) thick.

CHAPTER 3

Marzipan and Almond Paste

Before a cake can be iced, covering with either marzipan or almond paste is required. Marzipan is a manufactured paste and has a smoother texture than almond paste which is made from ground almonds, sugar and eggs.

Both products are suitable for covering cakes, but care must be taken when making almond paste to ensure that the paste is firm and oil from the almonds is not extracted by over-kneading. This could come through the icing and cause staining.

Almond paste and marzipan are expensive products and should be treated carefully. Hands must always be thoroughly washed before handling the paste and left-over paste should be well wrapped to avoid crusting. If it is placed in a polythene bag and tightly sealed to exclude air, it will keep for several weeks.

In addition to being the foundation for icing, almond paste and marzipan can be used for other purposes and surplus paste need never be wasted. Marzipan roses are a very pleasing and popular form of decoration, but almond paste does not make satisfactory flowers as it is too gritty. Fruits can be made with either almond paste or marzipan and arranged in boxes to make ideal gifts for occasions such as Easter, Christmas and the like.

I prefer to use marzipan rather than almond paste and use 'natural' marzipan which does not contain any artificial colouring. When colour is added (when making fruit or flowers) the use of 'natural' marzipan ensures that the required colour is obtained.

For those who wish to make almond paste, here is a recipe.

225 g (8 oz) icing sugar
225 g (8 oz) caster sugar
225 g (8 oz) ground almonds
Enough whole eggs or egg yolks to make a firm paste

Method

1. Sieve together all the dry ingredients.
2. Bind to a firm paste with beaten egg. A little rum or brandy may be added to give extra flavour.
3. Be careful not to knead the mixture and therefore express the oil from the almonds which would cause staining on the finished cake.
4. Wrap immediately to prevent the paste from crusting.

This mixture will weigh just over 675 g ($1\frac{1}{2}$ lb) and be enough to cover the top and sides of a 1·36 kg (3 lb) cake, ie. a 20 cm (8 in.) round fruit cake.

Apricot Purée

Before commencing to cover the cake with almond paste or marzipan it is coated with apricot purée. The purée is not applied merely to stick the paste to the cake, but to form an essential seal preventing fermentation taking place. For this reason, the purée is applied to the cake whilst boiling hot.

Purée is made by bringing slowly to the boil the contents of a 340 g (12 oz) jar of apricot jam together with the juice of half a lemon and a little water. It is difficult to give the exact quantity of water required because this depends on the consistency of the jam and some are thicker than others.

Start by adding a tablespoon of water and if it appears too thick, add further water until a soft but not runny consistency is obtained. Do not add more than three tablespoons of water as this would produce a purée that would be too thin to apply to the cake.

Boil this mixture *very gently* for five minutes. Sieve and discard any skins. Bring to the boil and simmer for a further five minutes before pouring into a clean jar. Keep covered until required.

Covering the Cake with Marzipan or Almond Paste

Before commencing to cover the cake, have ready to hand the following:

Marzipan or almond paste
Apricot purée
Cake board
Small pan
Palette knife
Icing sugar
Rolling pin
Small ruler
Tapemeasure
Small sharp knife
Clean, damp dishcloth
Turntable

Choose a clean work surface and check that the top of the cake is level; if not, trim with a sharp knife. In some cases it may be preferable to turn the cake over and use the bottom as the top to give a more even surface on which to apply the paste.

Selecting the Correct Cake Board

This is done by measuring the diameter of the top border drawing and using a board slightly larger than this measurement.

If the board is smaller than the top runout border, the finished cake will appear top heavy and the border can be easily damaged. Just how much larger a board should be is up to the individual to decide. I would place a 20 cm (8 in.) diameter cake with a border measuring 27 cm ($10\frac{1}{2}$ in.) across on a 31 cm (12 in.) board.

Do remember that borders vary in width, so it is not possible to say that all 20 cm (8 in.) diameter cakes will require a 31 cm (12 in.) board. Some will only require a 28 cm (11 in.) whilst others may require a 33 cm (13 in.) board.

Covering the Top

This procedure is the same for either marzipan or almond paste and either square or round cakes.

1. Bring a pan of purée slowly to the boil.
2. Sprinkle the working surface with a light dusting of icing sugar to prevent the paste from sticking.
3. Roll out approximately two thirds of the paste into a piece slightly larger than the top of the cake.
4. Take the cake to the pan of boiling purée and with the use of a palette knife thinly spread the top surface of the cake with purée. (Some people prefer to use a pastry brush for this purpose though I find a palette knife more suitable and easier to clean).
5. After making sure that the paste has not stuck to the working surface press the cake, puréed surface down, on to the paste and trim away any surplus paste with a sharp knife.
6. Press the paste to the side of the cake and make smooth with a palette knife.
7. Turn the cake the right way up and place in the centre of a silver (or gold) cake board.

Covering the Side

Round cake

1. Make sure the work surface is perfectly clean and free from any crumbs. Dust with a little icing sugar.
2. Measure the circumference of the cake with a tapemeasure. Form the remaining paste into a sausage shape and roll out a thin strip, a little longer than the required length.
3. Using a sharp knife trim the bottom of this strip. Measure the depth of the cake with a ruler and, using the point of the knife, mark the exact depth of the cake on the paste, at approximately 2.5 cm (1 in.) intervals.
4. Cut along the marks and, if necessary, trim both ends of the paste. This strip of paste should now be the exact height of the cake.
5. Roll up this strip like a bandage.
6. Spread the side of the cake with boiling purée.
7. Place the cake, still on the board, on the turntable.
8. Being careful not to stretch the paste, gently unroll the paste round the cake. The paste should not overlap where it meets and this can be avoided by trimming any surplus paste with a sharp knife.

Square cake

With a square cake the procedure is similar to the above, but to ensure that square corners are maintained, individual strips of paste are fitted to each side in turn.

If rounded corners are preferred, then the procedure is exactly the same as for a round cake.

For both round and square cakes it is not absolutely essential that the side strip is continuous. At first it might be found easier to make several short strips and piece them together on the cake. Ensure that they do not overlap otherwise a bulge will show on the iced cake.

After the side of the cake has been pasted ensure that the cake board is free from icing sugar or purée by wiping with a clean damp cloth.

If possible, leave for several days before applying the first coat of icing. This allows the paste to harden slightly, making it easier to work on. However, if time is short, a first coat of icing can be applied immediately, and in fact some people prefer to do so.

Once it has been pasted, do not wrap or place the cake in a tin. Allow it to dry out in a cool dry place.

CHAPTER 4

Marzipan/Almond Paste Fruits

Fruits will keep for several weeks but are best eaten within three weeks of being made. After this time, they become rather hard and tend to lose their flavour.

Strawberries are placed in small paper or foil cases to avoid the sugar coating being in contact with other fruits. It is not necessary (though it may be preferred) to use cases for all fruits, but if cases are not used, the box should be lined with a piece of clean greaseproof paper before the fruits are arranged.

Fruits may be made as large or as small as desired but should be of a uniform size. Each of the fruits illustrated in this chapter was made with a ball of marzipan the size of a large cherry.

Requirements

- Marzipan or almond paste
- A little egg white
- Granulated sugar
- Angelica
- Food colourings
- Food flavourings (if desired)
- Nutmeg grater
- Cocktail sticks
- Fine paint brush
- Small paper or foil cases
- Greaseproof paper
- Small sharp knife
- Small scissors
- Rolling pin
- Small polythene bags
- A little icing sugar
- Clean damp dishcloth
- Tray
- Boxes

Advance Preparation

Colour and Flavour the Paste

This can be done at least a week in advance and it is essential to do it several days before to ensure that the paste is not too sticky to handle.

The colours you use will depend on the fruits you wish to make. I have included nine fruits, but there are many more which you might like to try. Both liquid and paste colourings are suitable for using with almond paste or marzipan.

To add colour, take some marzipan or almond paste and ensure that the paste not being used is well wrapped in polythene to prevent crusting. After washing hands, add colouring to the paste with a cocktail stick and fold in with the fingers until the desired colour is reached and no streaks of colour are showing. Remember that the colour will darken a little as the paste dries, so add too little colouring rather than too much! If you wish to add the appropriate flavouring (and some people do not) then it should be added now. When the correct colour and flavour have been achieved, place immediately in a polythene bag and seal. Use a separate bag for each colour and store in a dry place until required.

In addition to the fruit a little green and a very small amount of brown paste will be required for leaves etc.

Strips of Angelica

Angelica will be required for stalks and can be prepared as much as a week in advance. Cut thin strips approximately 1 cm ($\frac{1}{4}$–$\frac{1}{2}$ in.) long and leave on a plate to become hard. This will make them much easier to insert into the paste.

Boxes (Fig. 7)

Boxes suitable for fruits can be collected and put on one side until required. These include notelet boxes and those which contained Christmas cards. Boxes with clear tops are particularly suitable, but any which previously contained products with a strong smell (ie. soap, cosmetics, etc.) must be avoided. It is also possible to buy boxes made especially for this purpose.

Fig. 7. Boxes.

Last Minute Preparation

1. Have to hand food colourings, a small sharp knife, a fine paint brush and a clean damp dish-cloth.
2. Place a sheet of greaseproof paper on a tray (fruits are left on the tray until firm; for several hours or preferably overnight).
3. Put the egg white in a small container and the granulated sugar in a bowl. (These are required only when making strawberries).
4. A fine grater is required (for use when making oranges and lemons).
5. Have ready a small polythene bag and several cocktail sticks. (Both will be needed when making leaves. Leaves are not made in advance as they become brittle).
6. A small rolling pin, icing sugar and scissors (which will be required only when making pineapples).
7. Have ready to hand a supply of small paper or foil cases.
8. Ensure that hands have been well washed before handling paste.

Cherry (Fig. 8)

Using the palms of your hands, roll a piece of red coloured paste into a ball. Place on greaseproof paper. Make a leaf, place it on top of the cherry and secure by stabbing with a strip of angelica.

Fig. 8. Cherry.

Leaf (Figs 9 and 10)

Roll a very small piece of paste into a ball. Roll to produce a cone shape (Fig. 9). (A little brown paste added to the green will give a mottled effect). Place cone inside a polythene bag and press flat using the forefinger. Without removing it from the bag draw the veins on the leaf with the end of a cocktail stick or the point of a knife (Fig. 10). Remove from the bag and pinch the wide part of the leaf so that it bends a little. Immediately attach to the fruit with a thin strip of angelica. If the leaf is allowed to dry before attaching to the fruit it will break when secured with the angelica.

Fig. 9. Cone, shown approximately four times normal size.

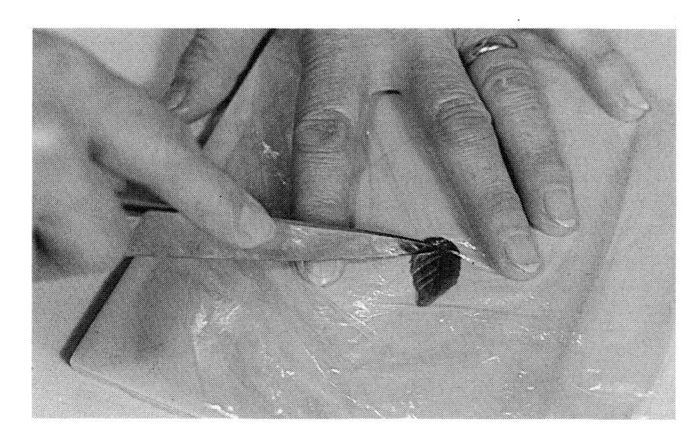

Fig. 10. Leaf.

Pear (Fig. 11)
Roll a piece of green paste into a ball. Roll the ball to produce a pear shape. Place a *very small* ball of brown paste at the base (thick end) of the pear and secure by spearing with the point of a cocktail stick. Secure a leaf on the top of the pear with a strip of angelica. Place on greaseproof paper.

When all pears have been made, brush with a little red or pink edible colouring.

Fig. 11. Pear.

Apple (Fig. 12)
Using pale green paste, proceed as with the cherry. Add a little red or pink colouring to give a blush to the apples.

Fig. 12. Apple.

Peach (Fig. 13)
Roll a piece of peach coloured paste into a ball. Using the side of a cocktail stick, make a vertical indentation in the peach. Secure a leaf and place on greaseproof paper. Use a little red or pink colouring on the finished product.

Fig. 13. Peach.

Strawberry (Fig. 14)
Roll a piece of red paste into a ball and then into a pear shape. Brush all over with egg white and drop into a bowl of sugar. Shake the bowl to ensure that the paste is well coated. Pick up the strawberry and shake off the surplus sugar. Place into a paper case; make a leaf and secure with a strip of angelica.

Fig. 14. Strawberry.

Orange (Fig. 15)
Roll a piece of orange coloured paste into a ball. Roll it around a grater to produce a rough 'peel' effect. Roll a *very thin* 'sausage' of brown paste with the palms of your hands and cut two strips approximately 1 cm ($\frac{1}{4}$–$\frac{1}{2}$ in.) long. Place these two pieces cross-wise on the top of the orange and secure by pushing the point of a cocktail stick through the centre. Place on greaseproof paper.

Fig. 15. Orange.

Lemon (Fig. 16)
Roll a piece of lemon coloured paste into a ball and then around a grater as with the orange. Pinch at both sides to form the shape of a lemon. Place a very small ball of brown paste at one end and secure with a cocktail stick. Place on greaseproof paper. A few dabs of green colouring can be brushed on the finished fruits.

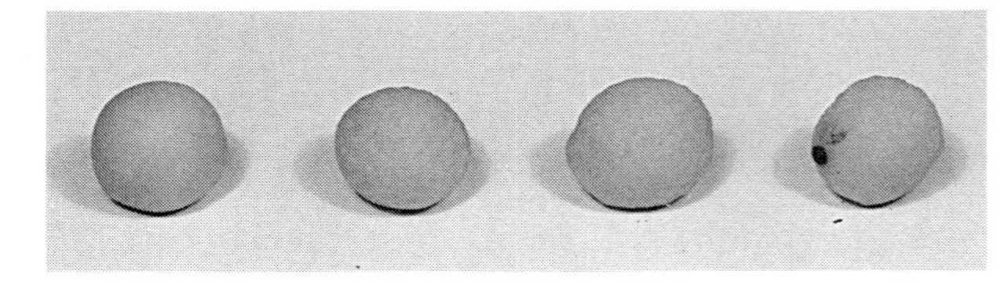

Fig. 16. Lemon.

Banana (Fig. 17)
Roll a piece of yellow coloured marzipan into a ball and then into a sausage shape. Bend a little to form a banana. Place a very small ball of brown coloured paste at one end and secure with a cocktail stick. Put on a sheet of greaseproof paper. When all the bananas have been made, add a dab of green colouring at each end and give ripeness by painting brown down the middle.

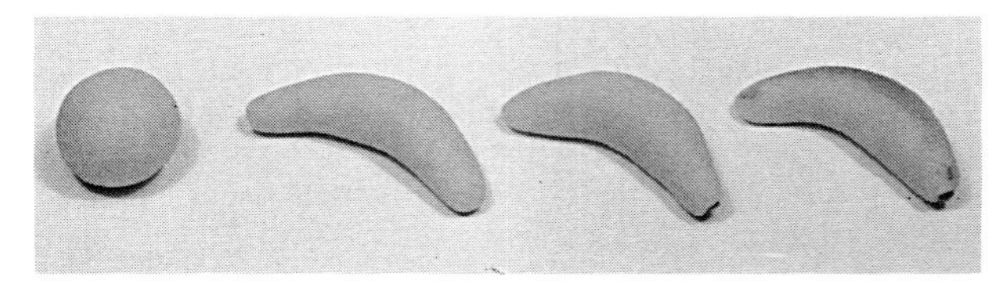

Fig. 17. Banana.

Pineapple (Fig. 18)
Roll a piece of yellow paste into a ball. Roll the middle of the ball to produce an oval shape. Place onto greaseproof paper and, with a sharp knife,

cut criss-cross lines on the surface of the pineapple. Roll out a thin strip of green coloured paste approximately 3 cm (1 in.) long and 1 cm ($\frac{1}{2}$ in.) wide. Using scissors, cut half way down the paste as shown. Roll up the strip like a bandage. Place a small ball of brown coloured paste at one end of the pineapple and secure with a cocktail stick. Use a cocktail stick to make a hole in the opposite end of the pineapple and place the green frond inside. Make sure it is secure by pressing with the cocktail stick. Place on greaseproof paper and use a little orange colouring to paint the finished fruits.

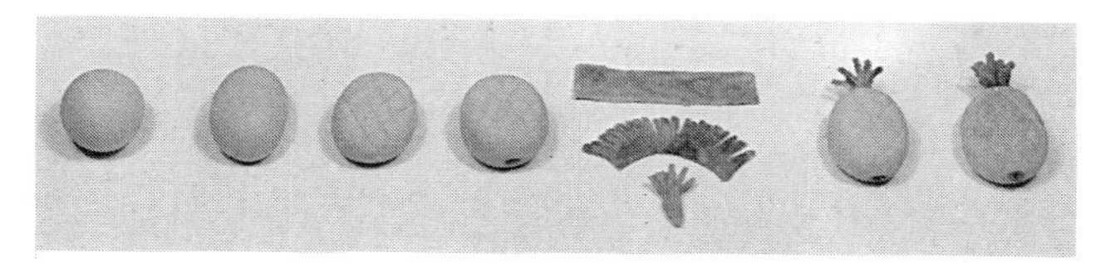

Fig. 18. Pineapple.

Fig. 19. Finished fruits.

CHAPTER 5

Marzipan Roses

Marzipan roses are featured in *The Art of Royal Icing* but their immense popularity and suitability as a decoration on royal iced cakes has prompted me to include them again in greater detail. Step by step illustrations show how to model large and small roses and further information gives several ways of arranging them on cakes.

Colouring the Marzipan

Marzipan is easier to handle if it is coloured several days in advance. If it is well wrapped in polythene to prevent crusting, it will remain manageable for up to two weeks.

If possible 'natural' marzipan should be used for modelling but if this is not available a good quality golden marzipan can be used. Almond paste is not suitable because of its coarse texture. The disadvantage of using golden marzipan is its basic yellow colour which combines with any colouring to give a secondary colour. It is impossible to achieve a delicate pink for example, and the delightful shades and harmony that can be achieved by colouring natural marzipan are unobtainable. The golden marzipan can be used effectively only with a white cake, when any distortion in the colours is not easily discernible.

Paste colouring is often used to avoid the marzipan becoming sticky, but in the majority of cases liquid colouring is quite acceptable unless a colour of great depth is required as with deep red roses.

Hands should be well washed before commencing to add colour to the paste with a clean cocktail stick. Fold in the colour using the fingers until the desired shade is achieved and the paste is no longer streaky. Immediately place in a polythene bag; exclude the air and seal. Leave in a cool dry place until required.

Only a small amount of marzipan which should be coloured green will be required when making leaves. If several shades of green are used for the leaves in one spray, the result is very realistic.

If an aerograph is available some of the finished leaves can be sprayed with a little brown colouring and this further improves their realistic appearance. As most people do not have such a gadget, however, the next best thing is to add a very small amount of brown marzipan to the green and fold it in until the green becomes speckled with brown.

It must be emphasized that although several different greens may be used in a spray of roses, the roses themselves must all be the same colour. For example, if a spray of roses is assembled on a shell pink cake, then the roses must also be shell pink and not French or rose pink! They may be a deeper shade of pink than the coating, but the colouring must be from the same bottle.

Shading the Mazipan

Roses of delicate colouring (pink, peach, yellow) look most attractive if the centre of the rose is darker than the outer petals with the colour gradually becoming lighter as each layer of petals is added. To achieve this gradation an amount of marzipan is coloured to give the darkest shade required for the centre of the rose. Cut the paste in half and place one piece in a polythene bag and seal. To the remaining paste, add an equal quantity of uncoloured paste and mix until no streaks of colour are showing. This colour will be paler than the first and can be used for the middle layer of petals. Cut this paste in half and, as before, place one piece in a bag and seal. Again add an equal quantity of uncoloured paste to the remaining piece of marzipan and mix to produce a very pale paste which will be used for the outer layer of petals. Place in a bag and seal. You will now have three equally graduated shades of the same colour.

Lustre Colours

A sheen can be given to a rose by dusting it with powdered 'lustre colour'. This edible powder, available in most colours, gives a slight sparkle to the finished rose. The powder is placed on a saucer and a little is brushed onto the edge of each petal with a clean dry paint brush. This should be done when the petal is assembled and before the marzipan is allowed to dry.

Making Roses

Large and small roses are made in almost exactly the same way. The exception is that large petals can be cupped around the thumb (Fig. 24) whilst small petals have to be gently bent between the two forefingers (Fig. 34). The size of a rose is governed by the size of the petal and each petal is formed from a ball of marzipan. To make a large rose each ball of marzipan will be about the size of a glacé cherry. A small rose will be made from balls of marzipan no larger than a small pea. Medium sized roses will be produced from somewhere between these two examples. The most important thing to remember is that whatever size of ball you start with, the same size must be continued throughout the rose. A good idea (learned from a student) is to make a ball of marzipan; place it in front of you and use it as a guide whilst modelling the rose.

Start by moulding a large rose and when you are satisfied with the result, practice making them smaller. Large petals are easier to handle. Very small roses tend to fall over until you become adept at handling small pieces of paste.

Points to Remember

1. A common fault when making roses is to make the petals too thin. The petal should be quite thick and only the very outer edge is thinned to give a delicate curl. Two problems can arise if the petal is too thin.

(a) The petal becomes too large for the cone and results in 'spare' paste underneath the rose.
(b) Because it is too thin, the petal 'falls away' from the centre of the rose. This also results in the rounded shape at the centre of the rose being lost.

2. Petals that are the correct thickness but have not been thinned at the outer edge will appear clumsy.

3. Failure to make a rose 'sit down' each time a petal has been assembled can result in

Fig. 20. Stages of making a rose.

the finished rose falling over and petals being damaged.

4 Only the final layer of petals is placed lower than the others. This will have the effect of opening-up the rose. Done too soon, however, it will spoil the shape of the rose.

5. Do not make every rose in full bloom. Buds and small roses will also be required; a spray of roses all the same shape will not look realistic.

6. Red roses should be a deep rich red. Roses in pale colours may be shaded and if so, the darkest colour is used for making the bud; a slightly lighter shade for the next layer and a very pale coloured paste for the final layer of petals. Petals can also be brushed with lustre colour and this can be applied to shaded or unshaded roses.

7. Roses should be left on the board until they are dry. If they have been made some time in advance, store in covered, but not airtight, containers. If an airtight container is used, the paste will become soft and the petals will collapse.

8. The instructions given for making a rose are designed to make it straightforward and easy to remember. You will find that eleven balls of paste are used to complete a full blown rose (Fig. 20) and the sequence is:

(i) Make into a cone.

(ii) Press into a petal and wrap around the cone.

(iii) Press into a petal, taper and place over the join of the previous one.

(iv) As (iii) but place *opposite* the previous petal. (These four stages form a bud and can be left as such. More petals may be added using a paler paste, if desired).

(v) (vi) (vii) Make as the previous petals but place around the bud in an interlocking fashion.

(This now produces a rose. If further petals are not required, make any adjustments with the fingers whilst the paste is still soft. A final layer of petals may now be added using a paler coloured paste if desired).

(viii) (ix) (x) (xi) Make in the same way as the previous petals and place around the rose in an interlocking fashion. In order to 'open' the rose, these petals are placed slightly lower than the previous ones.

When roses made this way are to your satisfaction, experiment by adding fewer or more petals. For example, add just one petal to a bud, or five petals (instead of four) to the final layer, letting the last one 'fall away' to give a natural effect.

Last Minute Preparation

Very little equipment is required to produce roses and leaves; a full list is as follows.

Sheet of greaseproof paper (only required when making leaves)
Coloured marzipan
Small polythene bag
Piece of board or a tray
Small sharp knife
Clean, damp dishcloth
(A fine paint brush and powdered lustre colour will be required if the petals are to be dusted).

Making a Large Rose

1. Remove a small amount of paste from the bag and, using the palms of both hands, roll into a ball about the size of a glacé cherry. Set down on a clean board or tray and use for reference.

2. Repeat the above procedure, checking that the ball of paste is similar in size to the one already on the board. Gently roll one side of the ball to form a cone and set it down on the board. This cone will form the centre of the rose (Fig. 21).

Fig. 21. Ball and cone.

3. Take out another piece of paste and roll into a ball the same size as before. Place the ball into a clean polythene bag and with the index finger on top of the bag, press into a disc of even thickness, about the depth of a 1p piece. Approximately three quarters of the circumference and about 4 mm ($\frac{1}{8}$ in.) from the edge of the marzipan is now pressed to a fine edge (Fig. 22). (This tapered section is used to form the curl of the petal. If it is tapered too much, the edge will fall back when assembled; too little and the petal will appear too thick.)

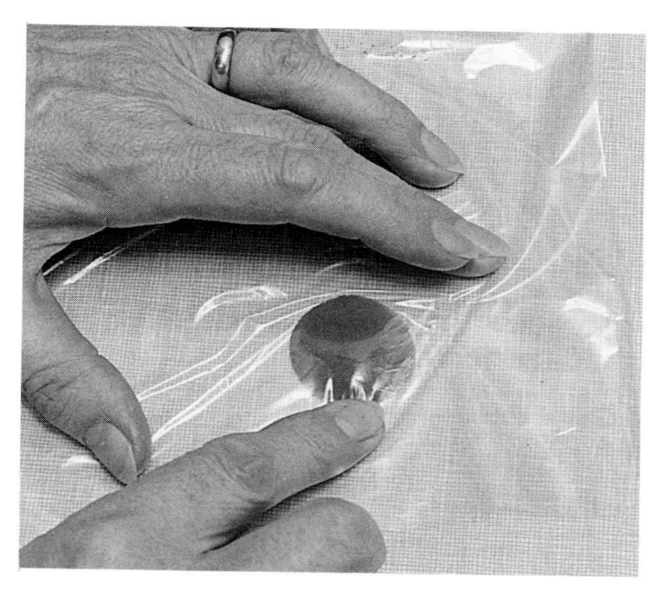

Fig. 22. Pressing petal in bag.

4. Remove petal from polythene bag and with the cone in one hand and the petal in the other, wrap the petal around the cone (Fig. 23). The tapered edge of the petal must be at the top of the cone and should form a small opening though not large enough for the point of the cone to be visible. Set the rose on the board.

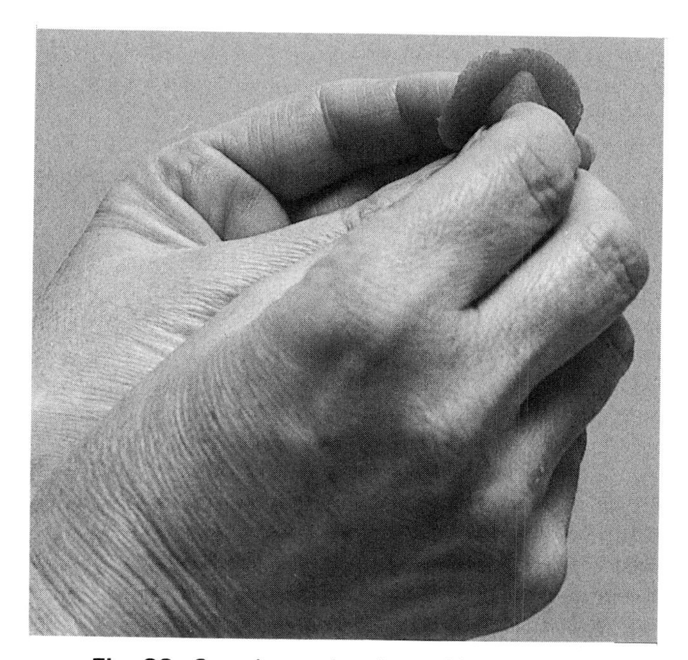

Fig. 23. Cone in one hand, petal in the other.

5. Make a ball and press out another petal, remembering to taper the edge as previously described. Take out of the bag and mould the petal around the thumb to produce a 'cupped' shape and gently curl back the top of the petal (thin edge) with the finger (Fig. 24). (Apart from the first petal which is wrapped around the cone,

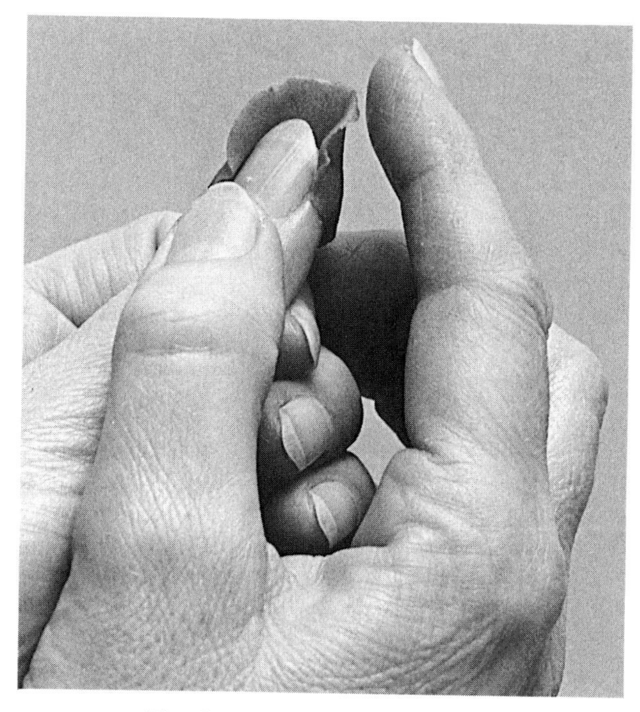

Fig. 24. Cupping a large petal.

all petals are 'cupped' and 'curled' before being placed on the rose.)

6. Place the petal on the rose covering the join of the previous petal. This petal should be placed *slightly* higher than the previous one and it will wrap around approximately three quarters of the rose (Fig. 25). Place on board.

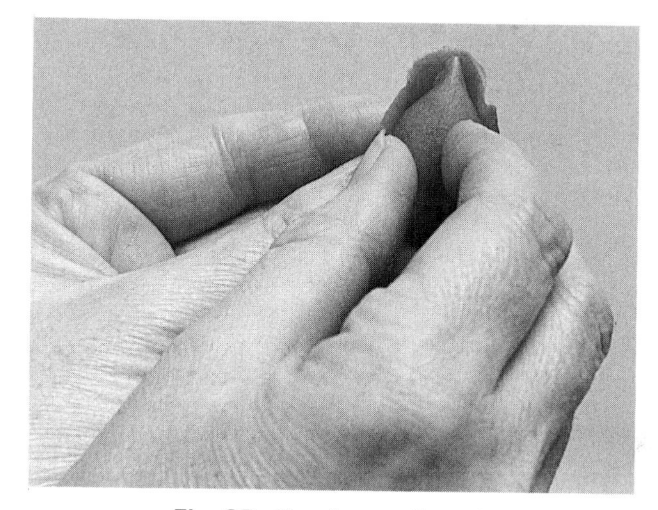

Fig. 25. First 'cupped' petal.

7. Make another petal in the same way and place opposite the previous one. When the rose is put on the board it will be seen that it now forms a bud (Fig. 26). Curl the petals a little more, if necessary, whilst the marzipan is still soft. If the rose is to be shaded, use a paler coloured paste for the following three petals.

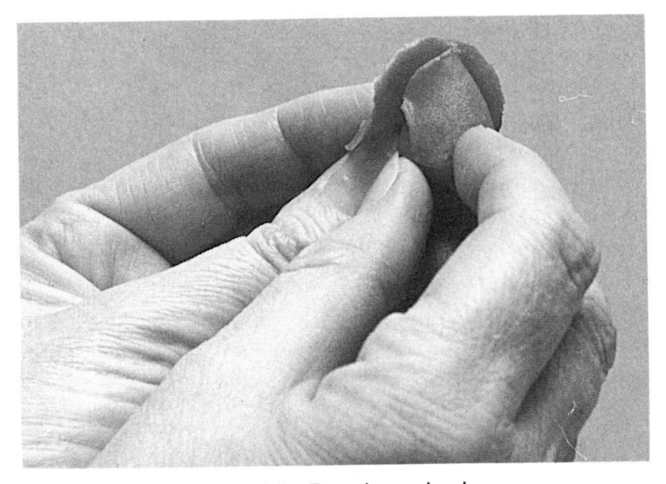

Fig. 26. Forming a bud.

8. Make another petal and place over a join making sure that this petal is the same height as the bud. This time only one side of the petal is placed to the bud; the other is left open in order that a second petal can be placed inside (Fig. 27). Place on board.

Fig. 27. 1st petal on bud.

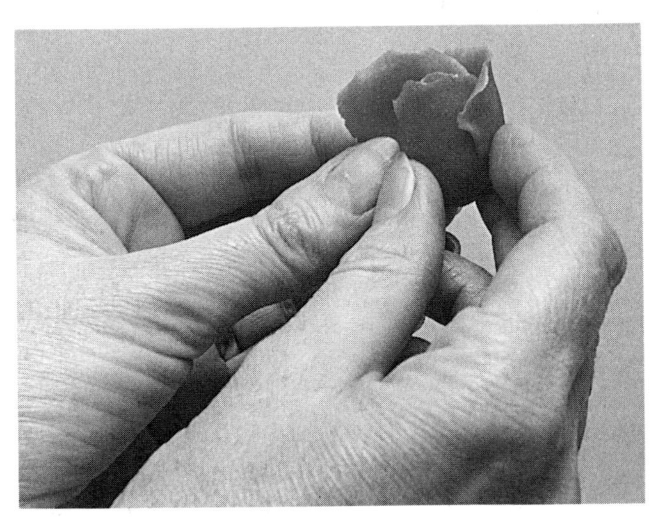

Fig. 28. 2nd petal on bud.

9. Make a second petal and place one half inside the first leaving the other half open (Fig. 28). Place on board.

10. A third petal is made to complete this layer and is placed inside the second. Press the petals gently to close and place the rose on the board (Fig. 29). If another layer of petals is required proceed as outlined in stage 11 and if they are to be shaded, use the palest paste.

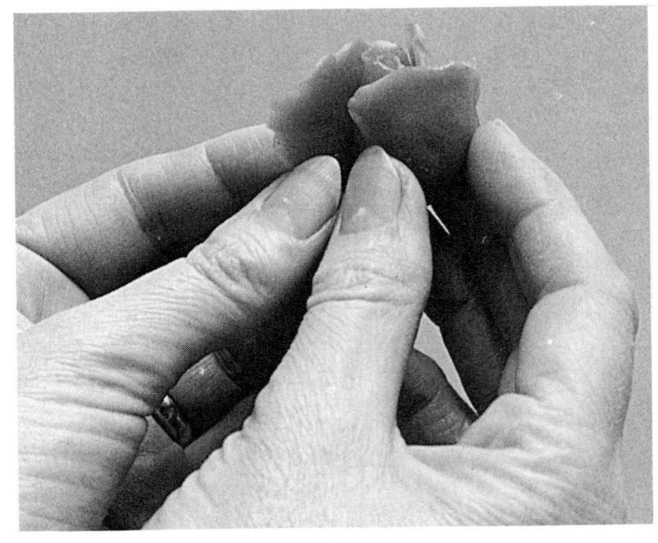

Fig. 29. 3rd petal on bud.

11. To produce a full-blown rose, four more petals are now added to the rose. Like the last three, each petal is inserted inside the previous one but placed in a *slightly* lower position to give an open effect. Each petal is separately moulded, assembled and placed on the board whilst another petal is being made. Any final adjustment to the curl of the petals must be made before the paste starts to harden (Figs 30–33).

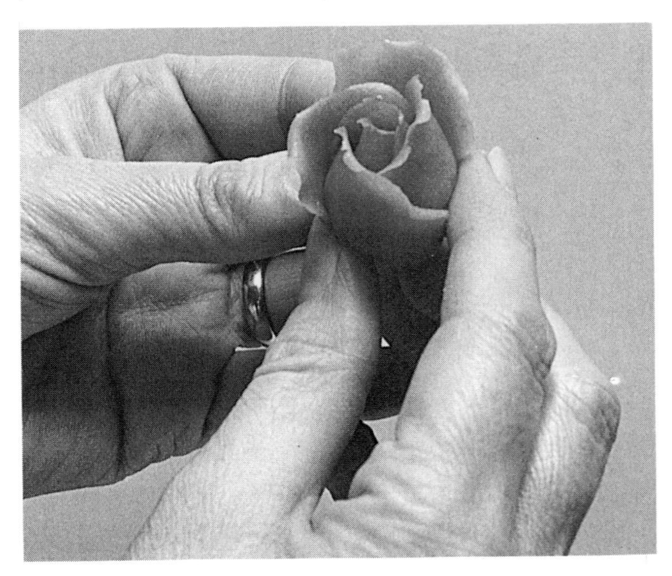

Fig. 30. 1st petal on rose.

Fig. 31. 2nd petal on rose.

Fig. 32. 3rd petal on rose.

Fig. 33. 4th petal on rose.

Making a Small Rose

Taking a small ball of marzipan (as small as you wish) proceed as with the instructions for making a large rose. Be sure to keep a ball of paste in front of you as a guide for size, and refer to it each time you make another ball.

When stage 5 is reached it will be found impossible to cup a very small petal around the thumb. Place it between the two fore fingers and bend slightly (Fig. 34). Curl the edge with the finger as previously described.

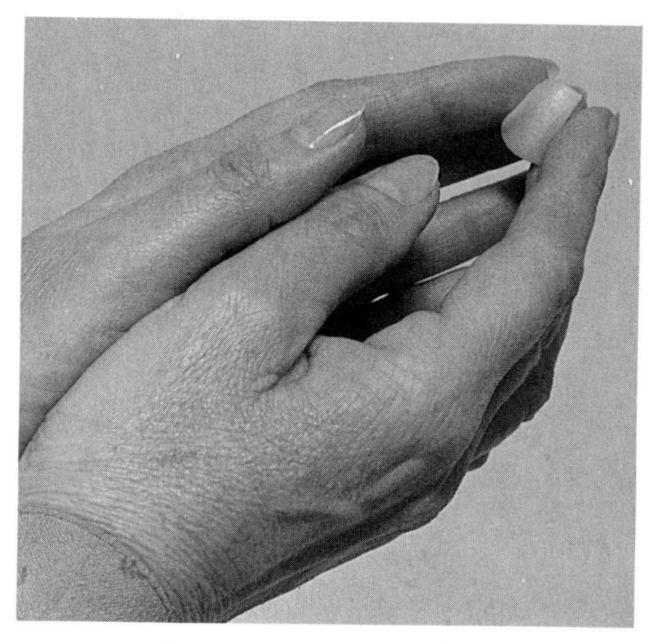

Fig. 34. Cupping small petal.

Apart from stage 5, all instructions given for a large rose also apply when making a small rose. Small roses tend to fall over easily and it is important that they are made to sit on the board each time they are placed down.

Leaves

These are simple and straightforward to make and are the same as shown on page 11 (Figs 9 and 10) for use with marzipan fruits.

Points to Remember

1. A leaf, like a rose, is made from a ball of marzipan.
2. A spray of roses should contain both large and small leaves. Make different sizes by adjusting the size of the ball of marzipan.

3. If you wish to produce a curved leaf, leave to dry over the handle of a spoon.

4. Make leaves at least several days in advance, leave until set and then store in non-airtight containers until required.

Making a Leaf

1. Roll a very small piece of green marzipan into a ball (or green marzipan that has had a small amount of brown marzipan added to it and folded until speckled with brown).

2. Form into a cone and place inside a polythene bag. Press out evenly, but not too thinly, to form the shape of a leaf.

3. Without removing it from the bag, draw the veins on the leaf with the point of a knife.

4. Remove from the bag and pinch the wide part of the leaf so that it bends a little.

5. Place onto a sheet of greaseproof paper until set.

Decorating with Roses

Before commencing to decorate a cake it must be stressed that both the coating and the flooding on the cake board must be completely dry.

Large Roses

Two ways of decorating a cake using large roses are outlined here.

Spray of roses (*Fig. 36*)
The most attractive way of using roses on a cake is to assemble them in a spray. This method is suitable for round or square cakes.

1. With dark brown coloured icing of piping consistency in a small bag fitted with a No. 0 tube, draw a line across the cake and pipe smaller lines branching off from the first as in Fig. 35. The tube is not lifted and should slightly scratch the surface of the cake.

2. Assemble the roses and place on the main stem arranging the leaves around them.

3. When satisfied with the positioning, secure to the cake with a little icing of the *same colour as the coating* (Fig. 36).

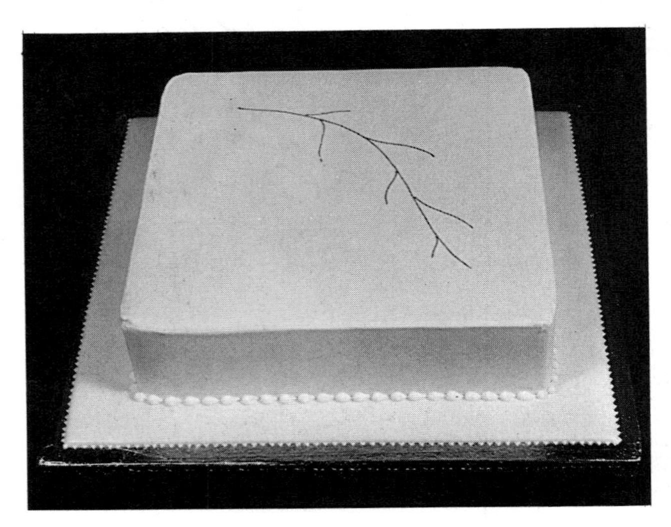

Fig. 35. Branches piped on cake.

Fig. 36. Roses assembled on cake — 23 cm (9 in.) cake on 30 cm (12 in.) board.

Single rose (*Fig. 37*)
A single rose placed on a plaque is a quick decoration for a cake. A very small cake will not require a plaque and the rose can be placed directly on to the cake and secured with a little icing of the same colour as the cake. The cake illustrated in Fig. 37 is 10 cm (4 in.) in diameter on an 15 cm (6 in.) board.

Fig. 37. Small cake with single rose — 10 cm (4 in.) cake on 15 cm (6 in.) board.

Small Roses

Here are three ways of using small roses:

Cluster (*Fig. 39*)
A cluster of small marzipan roses makes an unusual and edible decoration for the top tier of a wedding cake. As many roses are required, much time and patience is called for! The roses can, however, be made over a period of time and stored until you are ready to assemble them.

1. Have ready a plaque of the size and shape you require and a number of previously made roses and leaves.
2. With a little royal icing attach a cone of marzipan to the centre of the plaque (Fig. 38). The cone should be the same colour as the roses and its size will determine the shape of the finished cluster.

Fig. 38. Cone of marzipan on plaque.

3. Select a rose and secure it to the top of the cone with a little royal icing. Arrange and secure a single row of roses and leaves around the base of the cone. Fill in between the top and the base taking care not to damage the flowers. Any small spaces can be filled with rose buds, freshly made and still soft. Push these *gently* into position with the aid of a cocktail stick and carefully open the petals, if necessary, with a fine paint brush (Fig. 39).

Fig. 39. Cluster of roses on top tier of wedding cake — 13 cm (5 in.) cake on 20 cm (8 in.) board.

Spray (*Fig. 74*)
Small roses are assembled in exactly the same way as large roses but are only suitable for the top of a cake if they are first assembled on a plaque.

Fig. 40. Roses as side decoration — 20 cm (8 in.) cake on 30 cm (12 in.) board.

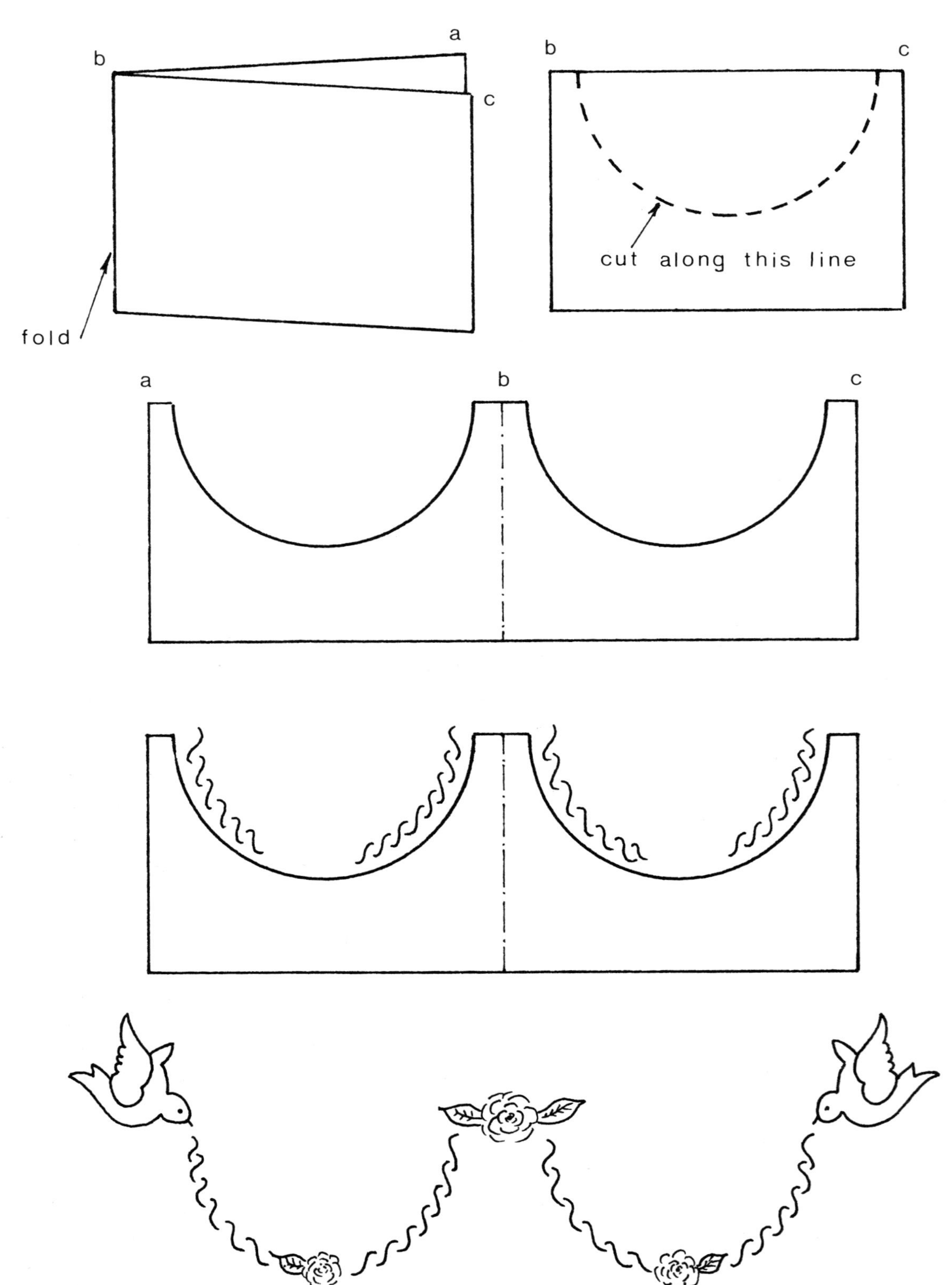

Fig. 41. Template for side of square cakes.

Side decoration (Fig. 40)

Small roses are suitable for decorating the sides of round and square cakes. Because they have much less weight than large roses they adhere quite easily to the sides of the cake.

The side decoration in Fig. 40 was carried out with the aid of a paper template which was placed on the side of the cake and held in position with three small bulbs of icing (Fig. 41). No. 1 linework was placed just above the template and the template removed before the remainder of the decoration was completed. This method ensures the curves are similar on all sides of the cake.

CHAPTER 6

Royal Icing

Ingredients for Royal Icing

Royal icing is a meringue made with egg whites and icing sugar. A little glycerine may be added to prevent it from setting very hard.

As when making meringues, it is essential that all utensils are scrupulously clean and absolutely grease-free. If fresh eggs are used they should be broken and the whites left in a clean, covered container for several hours before use.

Powdered hen albumen is preferable to fresh egg whites because it is much easier to use and there are no problems of what to do with the left-over yolks. Powdered albumen has, of course, to be reconstituted but once this has been achieved, the method of making royal icing is the same, whichever form of egg white is used.

Hen Albumen

This is easily obtainable by those in the trade, but sometimes hard to get otherwise. It is reconstituted by adding 300 ml (½ pint) of cold water to 45 g (1½ oz) of powdered albumen. After stirring well it should be covered and allowed to stand for at least two hours. Before use it should be stirred again and strained into a clean mixing bowl. This amount of liquid will take approximately 1·6 kg (3½ lb) icing sugar. For smaller amounts, use three tablespoons of cold water to one tablespoon of powdered albumen. Once re-constituted the albumen will keep fresh for about three days in a refrigerator.

Meri-White Royal Icing and Meringue Mix

This is a fortified albumen. It is available in shops, and instructions inside each box explain how to reconstitute the powder. Although I prefer to use hen albumen, I do find that this powder is much easier to use as it dissolves instantly and there is no need to leave it to stand before use.

Fresh Egg Whites

These should never be used straight from the refrigerator. They should be broken into a clean bowl, covered and left at room temperature for several hours. This will cause them to liquefy and make mixing much easier. Great care must be taken to ensure that none of the yolk gets into the egg white. The fat from the yolk will not only discolour the icing but make it impossible to obtain the right texture. Three egg whites will take approximately 450 g (1 lb) of icing sugar.

Icing Sugar

This should be free from lumps. Always reseal the packet and store in a dry place. Should even the smallest lump be present then the sugar must be sieved before use. The sieve, like all utensils, must be absolutely clean.

Glycerine

This is added to the icing required for coating and direct piping but *must not* be used in icing required for runout work. It is stirred into the icing at the end of the mixing and prevents it from setting too hard. I use one teaspoonful of glycerine to every 450 g (1 lb) of icing.

Making the Icing

Icing can be made in an electric mixer, though I prefer to mix by hand and so avoid the presence of too many air bubbles.

Method

1. Place egg whites into a clean basin and stir in enough icing sugar to give the appearance of unwhipped cream. This will take approximately half the total amount of the sugar.

2. Add a small quantity of the remaining sugar and *stir well*, but do not beat, for about two minutes. Continue in this way until the desired consistency is reached. Mixing will take between fifteen and twenty minutes but may be left at any stage providing it is covered with a damp cloth.

3. The correct consistency for coating and piping has been reached when the icing holds its shape when lifted from the bowl on a spatula.

4. Do not add glycerine to the icing if it is to be used for runout work. Should part be required for runout work, take out the required amount and place in a separate container. Add glycerine to the remainder to be used for piping or coating.

5. Colour may be added to the icing at this stage if required. Add very sparingly with the aid of a clean cocktail stick, remembering that the colour will darken slightly as it dries.

6. Scrape down the sides of the basin and cover with a clean damp cloth or place in a plastic basin with an airtight lid. There is no need for it to be kept in a refrigerator.

Coating the Cake

Obtaining a smooth, flat surface is the most important and possibly the most difficult procedure in icing a cake.

Preparation for Coating

Prepare the icing as described, or if it has been made previously, re-stir and add a little more sugar if necessary.

It is always advisable to have a little extra egg white available so that the icing may be diluted if required.

Before starting to coat the cake, have the following equipment to hand:

Turntable
Spatula
Palette knives
Straightedge
Side scraper
Clean, damp dishcloths

Icing the Top (Round or Square Cakes)

1. Place the cake board with the cake on the turntable. With the larger palette knife put a quantity of icing on top of the cake. Immediately re-cover the remainder of the icing in the bowl with a damp cloth to prevent crusting

2. With the palette knife, spread the icing evenly over the top of the cake and for a few moments paddle down the icing using a backwards and forwards movement. At the same time slowly rotate the turntable with the other hand. The paddling movement will help displace any air bubbles that may be in the icing. Should any icing spill down the sides of the cake, it can be removed easily when the top has been coated.

3. Holding the straightedge at the furthest edge of the cake and at an angle of about 45° to the surface, draw it across the cake towards yourself in one continuous movement. The surplus icing on the straightedge should be returned to the bowl and the straightedge wiped clean.

It is difficult, initially, to obtain a satisfactory surface, but a little practice produces remarkable improvement. Should the first attempt not be satisfactory, continue to paddle down the icing and re-scrape the top until a satisfactory coat is obtained. This process can be repeated for only about five minutes, otherwise the icing will start to crust and form hard particles. These cause lines in the coating and give an unsatisfactory finish. If this occurs, leave until the icing is dry and apply another coat.

4. Remove any icing from the side of the cake but, if this is dry, do not return it to the basin with the rest. It is far better to waste a little icing than to ruin the whole mix.

5. Leave the cake in a dry atmosphere to harden. It is inadvisable to leave iced cakes in a kitchen where they may be in contact with steam. Moisture absorbed into icing causes it to remain soft.

After about two hours the top should be dry enough to enable icing to be applied to the sides. It is possible to ice the top and sides together, but it is easier to do them in two stages.

Icing the Sides

Round cake

1. Make sure the cake is in the centre of the board and place it on the turntable. Spread icing on the side with a small palette knife. Paddle it down as on the top of the cake, slowly rotating the turntable at the same time.

2. Hold the side scraper at a slight angle (approximately 15°) against the side of the cake and with the other hand take hold of the board at the back of the cake near the scraper. Slowly revolve the turntable in one continuous movement until a circle has been completed. Lift the scraper away from the side of the cake. This will leave a take-off mark which initially will be very pronounced. With practice and after subsequent coats this mark will be less noticeable. (When decorating the cake, position the take off line at the back and, if possible, cover with some form of decoration.)

As when icing the top; several attempts at obtaining a smooth coat can be made before the icing starts to dry.

Some icing may have found its way on to the top of the cake. As the top coat is dry it is an easy matter to remove the surplus with a palette knife, leaving the top coat unharmed. Any icing on the cake board should be removed before leaving the cake to dry. Apart from being unsightly, hard icing prevents the smooth movement of the side scraper, resulting in an uneven finish.

Square cake

With one exception, the procedure for coating the sides of a square cake is the same as for a round.

Instead of revolving the turntable and icing the side in one continuous movement, each side is coated individually. To obtain neat corners, alternate sides are coated and left to dry before the others are done. Surplus icing on top of the cake and adhering to the bottom board should be removed before it is dry.

It is not absolutely necessary to have a turntable when coating a square cake.

Subsequent Coats

One coat of icing is never enough. Two are absolutely essential and three preferable. The beginner may find it entails four or even five coats before a satisfactory result is obtained.

Second coat

When the first coat is completely dry, remove any rough edges and the take off line with the aid of a sharp knife or fine sandpaper. Make sure there are no loose particles of icing on the cake which would spoil the next coat of icing.

Using icing of a slightly softer consistency, apply another coat in exactly the same way as the first. Icing left over from the previous day that has been covered with a damp cloth will probably be the right consistency. It will, however, need stirring before use.

Third or final coat

When the cake has been coated to satisfaction it will require a final coat of icing. This should be softer than the second coat, about the consistency of slightly whipped cream. Icing mixed the previous day is better for this purpose as any air bubbles will have dispersed.

The final coat is applied in exactly the same way as the others, but with a little more pressure applied to the straightedge and scraper. Most of the mixture is taken off the cake, leaving a thin, smooth film of icing. Because the consistency of the icing is thinner and the pressure applied greater, the take-off line should be barely visible. If it does require attention, use a sharp knife and smooth over the mark with a clean, damp paintbrush. A knife or sandpaper must not be used anywhere else on the final coat of icing as they would leave marks on the surface of the cake.

Leave the cake in a dry atmosphere. Do not put in a tin but, when dry, cover if necessary with a clean piece of white tissue paper.

The Piping Bag

I use bags made from greaseproof paper, though some people prefer to use silicone paper. Two sizes of bags will be required; the larger for flooding large areas (runout borders, cake boards etc.) and when using a large tube (No. 3 upwards). The smaller bags are used for runout figure work, small areas of flooding and when using tubes Nos 0, 1 and 2.

Do not buy a roll of paper for making bags but a packet with sheets measuring 38 cm (15 in.) by 25 cm (10 in.).

Cutting Paper for a Large Bag

Fold a piece of greaseproof paper 38 cm (15 in.) by 25 cm (10 in.) *diagonally* and cut along the fold using a sharp knife. This sheet of paper will make *two large bags* (Fig. 42).

Cutting Paper for a Small Bag

Fold a piece of paper 38 cm (15 in.) by 25 cm (10 in.) *in half* and cut with a sharp knife. This will give two pieces each measuring 25 cm (10 in.) by 19 cm (7½ in.). Proceed to fold and cut these

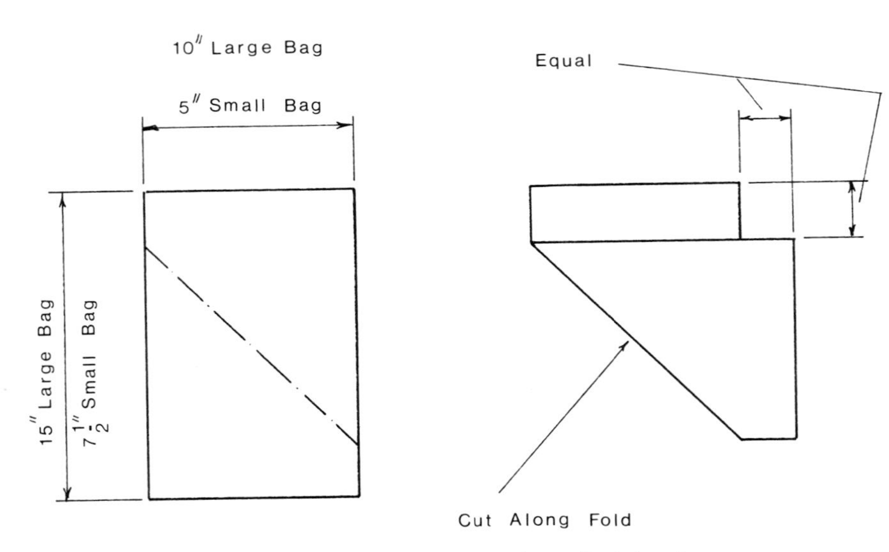

Fig. 42. Cutting paper for large and small piping bags.

papers in exactly the same way as for a large bag. This sheet of paper will make *four small bags* (Fig. 42).

Making a Piping Bag

The procedure is the same for large or small bags.

1. Hold the paper with the short straight edge between the thumb and fingers of the right hand (Fig. 43(a)).
2. Take the point in the left hand, and wrap it over and around the right hand (Fig. 43(b)).
3. Move the right hand backwards and forwards, at the same time pulling with the left hand until a sharp point is formed (Fig. 43(c)).
4. Place the left hand on top of the right and take hold of the bag, firmly, with the left hand (Fig. 43(d)).
5. Tuck in flap and make secure.

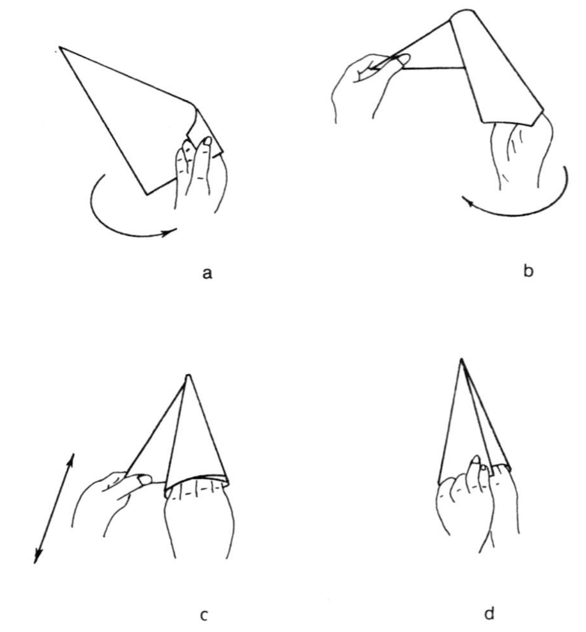

Fig. 43. Making a piping bag.

CHAPTER 7

Direct Piping

This is the term I use to describe work executed on the cake itself as opposed to runout work which, initially, is piped on to waxed paper and when dry is lifted onto the cake.

When piping on a cake, the coating must be completely dry and any mistakes carefully removed without damaging the icing.

Piping Exercises

All work to be carried out on a cake should be planned and practised beforehand.

Requirements

Bowl of icing, prepared to the consistency of first coating
Piping tubes Nos 3, 2, 1 and 0
Piping bags; large and small
A practice board (a piece of formica is ideal, especially if dark in colour)
Small palette knife
Scissors
Fine paint brush
Clean, damp dishcloths

Filling the Piping Bag

To fill a piping bag, first cut 1 cm ($\frac{1}{2}$ in.) from the bottom of the bag with sharp scissors. Place a tube in the bag and use a small palette knife to transfer the icing from the bowl. Fold over the top of the bag and roll down neatly until the icing is reached. Never be tempted to fill the bag more than half full. Re-cover the bowl and place the point of the tube in a clean, damp dishcloth.

Occasionally a No. 0 tube will become blocked in spite of the icing sugar having been sieved. This can be avoided by using the following method:

1. Place a No. 0 tube in a small piping bag.
2. Put a small amount of icing in a piece of nylon stocking.
3. Place the piece of stocking, containing the icing, inside the piping bag.
4. Take hold of the bag in the left hand, and hold it closed with the finger and thumb. Pull out the stocking with the right hand, leaving the sieved icing in the bag (Fig. 44).

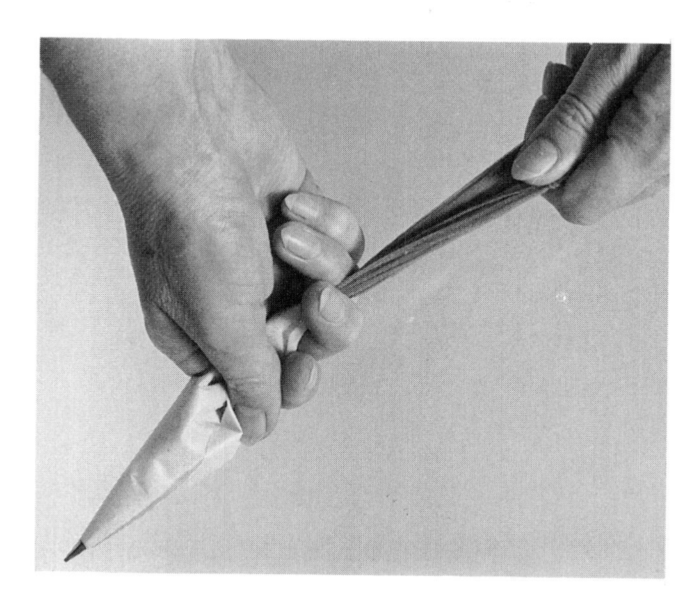

Fig. 44. Icing sieved in piece of nylon.

Holding the Piping Bag
Holding a piping bag correctly is important. There are two ways and both are fully explained and illustrated in *The Art of Royal Icing*.

When a small bag is required
For small areas of flooding, fine linework, bulbs, filigree and small shells; very little pressure is required and the bag is held in the right hand between the thumb and first two fingers. The left hand assists in holding it steady.

When a large bag is required
For large areas of flooding and shell borders, greater pressure is exerted and the bag is held in the palm of the right hand and steadied by the left hand.

Linework
Linework for the top of the cake (inside the runout border); on the sides of the cake (underneath the runout border) and on the cake board (prior to flooding) is carried out using the *touch*, *lift* and *place* method.

The bag is not pressed until the tube has made contact with the surface of the cake or board. How high the tube is lifted depends upon the length of icing required.

1. *Touch* the surface with the tube at the same time lightly pressing the bag. As the icing starts to flow –
2. *Lift* the tube from the surface. When the icing is the required length, stop pressing and –
3. *Place* the icing down on the surface.

Directly piped lettering and outlines for runout work are achieved by using this method.

Shells
Shells are piped with tubes of all sizes. Those piped with a No. 2 or No. 3 tube are for piped border work. A large bag will be required and is held in the palm of the hand.

Very small shells, such as those used for edging 'curly' corner pieces will be made with a No. 1 tube in a small bag held between the thumb and first two fingers.

Regardless of the size of the shell, the action is the same. The tube is placed on the surface and the bag is kept steady once piping has commenced. The action is to *press* firmly whilst *lifting* the bag slightly. When the required size of shell has been formed, *stop* pressing and *pull* away whilst making contact with the surface again. Any take-off 'tails' will be covered by the next shell or can be flattened with the aid of a fine paint brush.

When piped quickly, this action becomes one of *push* and *pull*.

Filigree
Filigree is always piped with a No. 0 tube in a small bag. The tube is not lifted and almost comes into contact with the surface of the cake or (in the case of runouts) the waxed paper. Filigree work placed in a collar must frequently touch the outer linework otherwise it will collapse when lifted from the waxed paper.

Small bulbs
Small bulbs are made with a No. 0 tube in a small bag. The method for these is *press*, *stop* and *lift*. The bag is pressed firmly, but once the icing has appeared the pressure is stopped before the tube is lifted away. If the pressure is not stopped a 'tail' of icing will be left when the tube is lifted.

Bulbs are used to edge the flooding on cake boards. They are also used to edge plaques and runout borders and to form small flowers.

Small bells and leaves
Small bells and leaves are formed from bulbs made with a No. 0 tube, deliberately leaving a 'tail'. Veins on the leaves are made with a fine paint brush used whilst the icing is still soft.

Bows
Bows are piped with a No. 0 or No. 1 tube in a small bag using the *touch*, *lift* and *place* method. A small bow will only require that the tube is lifted very slightly.

Stems
Stems are piped with a No. 0 tube but here the tube touches and slightly scratches the surface on the cake (Fig. 35).

Figure 45 shows examples of the direct piping covered in this chapter.

Combining Direct Piping with Runout Work
By themselves, directly piped designs make quick and effective side decorations. Piped designs combined with runouts blend to give many varied designs (Fig. 46).

Fig. 45. Piping exercises with tubes Nos. 2, 1 and 0.

A runout plaque carrying a monogram for the top of a cake is made more attractive with the inclusion of direct piping made with a No. 0 tube. This work is the background and should not detract from the initials. Because of this, the piping is often carried out in the same colour as the plaque (Fig. 73).

The technique is firstly, to scratch the decorative design lightly onto the plaque; secondly, to place the previously made monogram; and finally to complete the decoration with a No. 0 tube (Fig. 47). (If the decoration were to be piped before the monogram is placed, the icing would harden and the monogram would not lie flat.)

Flooding the Cake Board

This is carried out after the coating of the cake has been completed. There are several ways of flooding the board and they require icing of piping consistency; a small bag fitted with a No. 1 tube and several large bags of softened icing.

Points to Remember

1. All bottom borders are flooded *before* top borders are assembled.
2. Icing containing glycerine may be used.
3. Estimate how much softened icing you will require and fill an extra bag as the border must be completed before the icing crusts.
4. Make sure there are no breaks in the outline through which softened icing might flow.
5. Before commencing, test a small amount of the softened icing and if it is not correct, adjust the consistency. It is easier to dilute icing that is too thick than to add sugar to icing that is too thin.
6. Icing should be flooded between the cake and the outline. Flooding on the outline may cause linework to break.
7. If the flooding is to be left without filigree work, it should be free from air bubbles. To eliminate bubbles, icing should be mixed and diluted several hours before use. It will still require to be stirred; gently but well, prior to being placed in the bags.
8. Leave to dry for several hours on a *flat* surface.
9. Where the design on the board is the same shape as the border (Figs 104 and 106) care must be taken that the border, when assembled, is directly in line with it.

Direct Piping For The Side Of A Round Cake

Direct Piping Plus Runout Bells & Hearts

Direct Piping For The Sides Of A Square Cake

Direct Piping Plus Runout Vase & Butterfly

Fig. 46. Designs for sides of cakes (round and square).

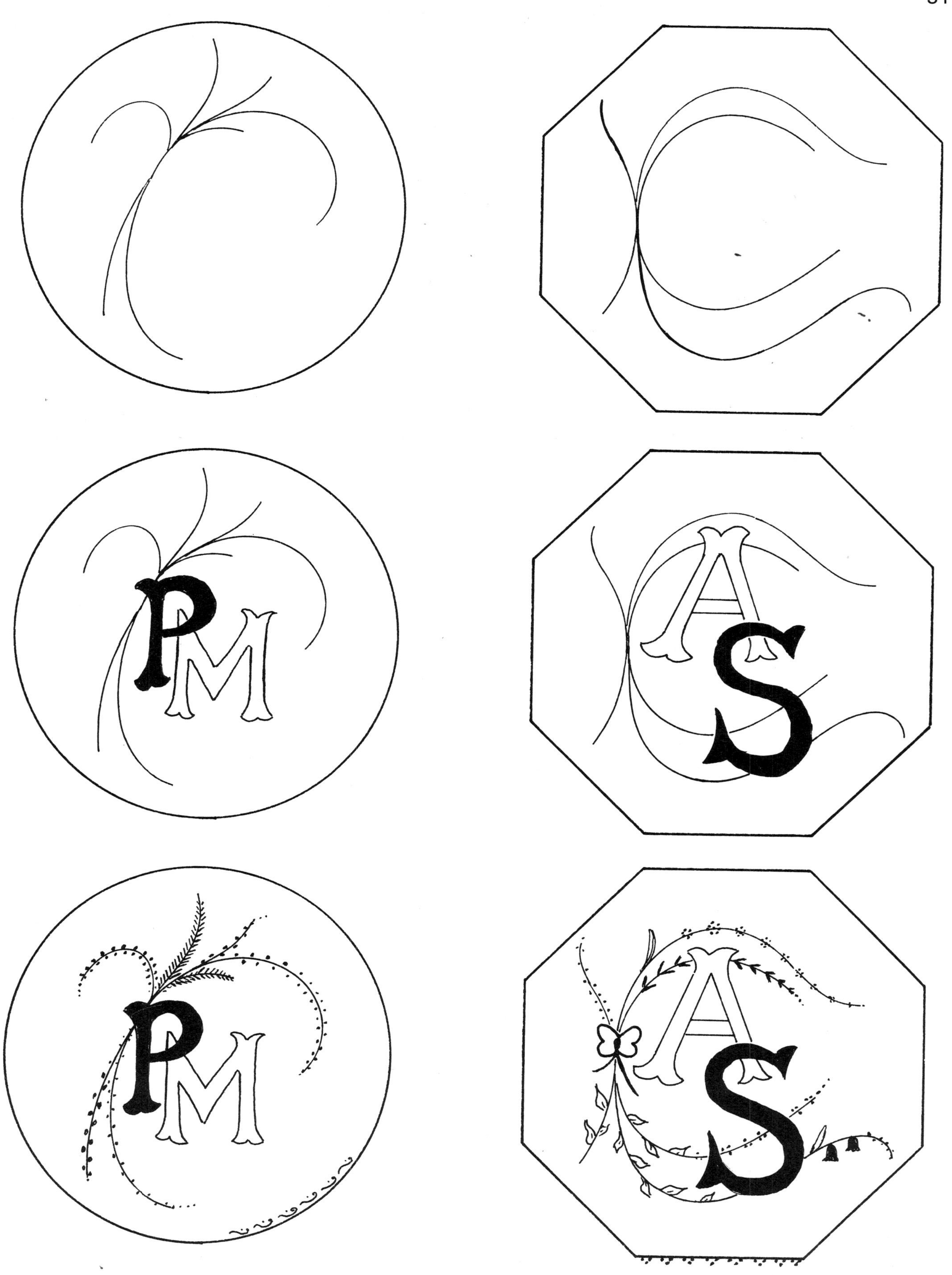

Fig. 47. Direct piping combined with monograms.

Method

Flooding the cake board, plain (round or square)

1. Dilute a quantity of icing with egg white or albumen, stirring gently until the correct consistency is reached. The icing should resemble thick cream and not leave the mark of the spatula when lifted from the bowl.

2. Half-fill several bags with this mixture.

3. Pipe a line with a No. 1 tube (*touch, lift* and *place*) inside the edge of the cake board, leaving room for an edging to be piped after the icing has dried (see step 5).

4. Take a bag of icing; cut a hole about the size of a No. 3 tube and test the consistency. Flood between the cake and the linework, using a fine paint brush to assist in smoothing the icing. Do not flood the whole board, but a small section about 5 cm (2 in.) wide.

Before this icing has been allowed to crust, continue flooding similar sections on alternate sides of the first one – ie. the left-hand side and then the right-hand side. By progressing around the board in this manner the icing will remain fluid and no join will be visible where the final sections meet. Leave to dry.

5. The border may be left plain or piped with filigree using a No. 0 tube. The edging is piped with a No. 0 tube also.

Flooding round cake board to match continuous runout border

To ensure the design on the board matches the top border a template is made from the original border drawing. This is achieved by cutting round the outer edge of the design and making a hole in the centre large enough to place over the cake as in Fig. 48.

Fig. 48. Template for flooding round board.

The template, thus formed, is placed over the cake; positioned on the board and, with a No. 1 tube, a line piped well to the outside of the template. This will allow the template to be removed without damaging the linework and will make the bottom border slightly larger than the top. Flood as directed for plain borders.

Flooding square cake board to match continuous runout border

When flooding a square board it may be found easier to cut the template into four pieces and place it on the board as shown in Fig. 49.

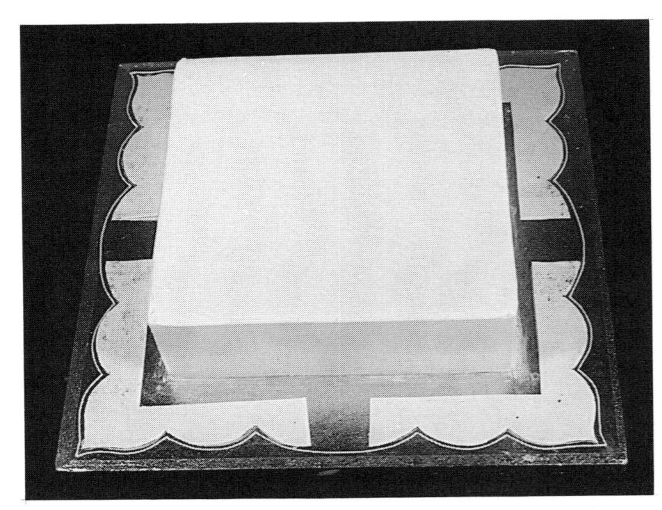

Fig. 49. Template for flooding square board.

Flooding round cake board to match separate border pieces

With a No. 1 tube, mark the top outside edge of the cake to indicate where the border pieces will be placed. Pipe small marks at the bottom of the cake in line with those on the top. Using a border piece drawing as a template, place the template in position on the board between the marks and pipe around it with a No. 1 tube. Flood as directed for plain borders.

Flooding square board to match separate border pieces

Using a border piece drawing as a template, place the template in position on the board and pipe around it with a No. 1 tube. Flood as directed for plain borders.

It will be seen that only the shape of the outer edge of the top border (or border pieces) has been used. If the entire border is to be repeated on the cake board, then the design will have to be cut out and outlined. See wedding cake Fig. 106.

CHAPTER 8

Runouts

The term 'runout' applies to all work that is piped on to waxed paper and allowed to dry before being transferred to the cake.

Runout work is, in my opinion, the most enjoyable and rewarding aspect of royal icing. It is also where individual style is developed, for no two people produce identical runouts. For instance; give a class a figure drawing and each student will interpret it differently.

The figure drawings included in this book can be carried out in many different ways. Some details may be added and others left out. Border work can be altered by changing filigree for linework or adding flowers or numerals.

Those already familiar with runout work will find the consistency I use for flooding runout figures much thicker than generally taught. This produces fatter figures which do not require to be outlined. It also means they are less fragile (Fig. 50).

Fig. 50. Christening cake — 18 cm (7 in.) cake on 28 cm (11 in.) board.

I have also included drawings for flat runout figures. These require outlining and the icing used for flooding is of a thinner consistency. Because they are not realistic, they are mainly used for children's cakes.

Drawings

The main feature of this book is the inclusion of working drawings suitable for runout work. In order to indicate the way in which they are piped, I have included drawings of additional stages required to complete the runouts.

Working Drawings

In practice this will be the only drawing required. As many as required can be traced, placed on boards and covered with waxed paper. After the runouts have been completed, the tracings are kept for future use.

Intermediate stage

These drawings are an indication that an additional stage is required to complete the runout. Some figures require extra flooding to that shown on the working drawings (eg. veil and bouquet for bride). Other figures might require decoration which can not take place until the runout is on the cake (eg. plaited hair, flowers in hand).

Final stage

Where a figure requires some form of decoration, this is indicated on this drawing.

Chapters 9–12 contain many working drawings for figure, plaque and border work that can be traced from the book.

Points to Remember For All Runout Work

1. Icing to be used for runouts must *not* contain glycerine.
2. Experiments show that large runouts made with fresh egg whites do not always dry out sufficiently, although small runouts are quite satisfactory.
3. Freshly made icing is required. It must not be more than two days old and mixed to piping/first coating consistency.
4. Piping tubes are not usually used for flooding.
5. Where required, outlines are piped with a No. 1 tube.
6. If colour is used, it should be added, with a clean cocktail stick, before the icing is diluted.
7. When using coloured icing for flooding, always mix more than you think you will require. Colours dry darker and are almost impossible to match perfectly. Half a pair of trousers, or part of a dress of a slightly different shade will be very noticeable when the figure is completed.
8. Icing required for flooding borders or plaques should be diluted several hours in advance to allow air bubbles to disperse. It must be stirred again, gently but well, before being placed in bags.
9. Bags of icing that do not contain tubes should not be more than half full and never be placed in a damp cloth.
10. Correct consistency is essential and can be achieved with a little practice. Three slightly differing consistencies are used for runout work.

 (a) Icing that will gently flow (outlined figures).
 (b) Icing that is soft but not runny (lettering, plaques, borders).
 (c) Icing that is as thick as possible in order to stay where required whilst soft enough to lose brush marks (runout figure piping).

11. Icing is always kept covered with a damp cloth or in Tupperware containers with air-tight lids.
12. Soft icing in bags should be checked for consistency before being used.
13. Bags of icing should not be held when not in use as the heat from your hand will alter the consistency.
14. A damp (but not wet) paintbrush is essential for assisting the flow of icing.
15. The basic colour of a runout border should be the same as the cake coating.
16. Large bags are only used when flooding cake boards, plaques and border work.
17. Flooding an area before the adjoining one has crusted will result in the areas merging.
18. When flooding, do not use all the icing in the bag as the heat from your hand will have caused it gradually to thicken.
19. Work should be done in a dry atmosphere; a steaming kettle may prevent a runout from drying.
20. Direct heat from a lamp is essential to assist the drying; prevent sinking and ensure a pleasing gloss to the finished runout.
21. Complete drying can take anything from a few hours to two days depending upon the size of the runout and the drying temperature. If a drying cabinet is not available, an airing cupboard will do. It is important that they be left in a dry, preferably warm place as moisture will cause colour to run and runouts to remain soft.
22. Additional piping and painting on runouts must not be done until the runout is completely dry.
23. When completely dry, plaques are stored, still attached to their waxed paper, in single layers. Runout borders and fragile figures are stored attached to the board on which they were piped.
24. After carefully removing the waxed paper, runouts are attached to the cake with a little icing of piping consistency.
25. Drawings should be kept flat and stored for future use.

Requirements

Icing mixed to piping consistency
Extra egg white or diluted albumen
Piping bags (small for figures; large for plaques and border work)
Fine paint brush
Thin waxed paper
Pieces of board, tiles, perspex etc.
Colouring
Cocktail sticks
Anglepoise lamp
Small palette knife
Clean dishcloths
Scissors
Masking tape
Working drawings
Pencil
No. 0 and No. 1 tubes

Advance Preparation

Advance preparation is very important and, in most instances, can start days, weeks or even months beforehand!

1. Draw or trace the work you intend to carry out on to pieces of greaseproof paper. Greaseproof paper is used because it is thin, transparent and will lie flat under the waxed paper.

2. Allow for breakages by tracing more than required. Surplus runouts will keep indefinitely if stored in boxes in single layers.

3. Place tracings on pieces of Perspex, glass, tiles or any material that is flat and will not bend. Avoid having more than one or two drawings on each board. Runouts are less likely to be damaged if well spaced. Using several boards enables some to be crusting under the lamp whilst others are being piped.

4. Place a separate piece of waxed paper, shiny side up, over the greaseproof paper containing each traced runout drawing and secure with masking tape at each corner. (The only exception to this procedure is that a group of very small figures such as lambs, small numerals, butterflies etc. may be piped together on one sheet.) The use of individual pieces of paper allows for easy removal, should any mistakes occur, without spoiling other figures.

5. Stack the boards, place several heavy books on top and leave as long as possible to ensure that the tracings and the waxed paper will be absolutely flat when the work commences.

CHAPTER 9

Piped Runouts

Piped runouts are those which do not require to be flooded. Most 'curly' corner pieces come into this category and they are included in this chapter.

Flower

Only *one* tracing is used for each size of petal, regardless of the number of petals required.

Several curved dishes (fruit saucers are ideal) will be required to use as formers to dry the petals to shape. Strips of fine waxed paper, not much larger than the drawing, will also be required.

Very little icing, which must be of piping consistency, and also a small bag fitted with a No. 1 tube are required.

1. Place the tracing on a board and stick down with tape.
2. Place a piece of waxed paper over the tracing and secure with a small bulb of icing.
3. Outline with a No. 1 tube and immediately remove the waxed paper from the drawing; place in the dish and secure with two small bulbs of icing.

Fig. 51. Petals placed in dish.

These petals are very fragile, though after a little experience surprisingly few are broken. As they are very quick to pipe, I suggest making double the amount required. Drying with a lamp is not necessary as the petals dry almost instantly and are not removed from the dish until assembled (Fig. 51). This is achieved by gently lifting from the waxed paper with a cocktail stick.

To Assemble Flower

As this flower is very delicate to handle (though will not break once it has been assembled) it is preferable to place on a small plaque rather than directly on to the cake.

1. Place a bulb of icing in the centre of the plaque and equally space eight petals in the icing. It is not necessary to use foam or cotton wool to keep them in an upright position as due to their light weight they are self supporting.
2. If one layer is required, leave for a few moments to set before piping a bulb of slightly softened icing in the centre.
3. If two or three layers are required proceed as before using a smaller drawing for each layer of petals, and place the second layer of petals *between* the first. Finish with a bulb of icing in the centre of the flower.

This flower looks most attractive if the petals are tipped with silver. Silver paint for this purpose is non-toxic but not edible and should only be used when this can be explained to the recipient of the cake.

The shape of these petals can be outlined and flooded to make a flower with solid petals. A lovely sheen will result if they are dusted with edible lustre colour when dry.

A layer of solid petals can be used with one, or more, layers of open petals.

Instruction for flooding petals can be found in *The Art of Royal Icing*.

In addition to the working drawing shown in Fig. 53, this flower is shown on the cake in Figure 101.

Butterfly

Butterflies are fragile, so ensure against breakages by tracing double the number required. Afterwards the tracings can be placed in envelopes and kept for further use.

1. Place tracings on a board, cover with a piece of waxed paper, shiny side up, and stick the corners down with masking tape.
2. Pipe over the design using a No. 1 or if preferred, a No. 0 tube (Fig. 52).
3. Leave to dry and do not remove from the waxed paper until required. The easiest way to do this is to lift gently from the paper with a small artist's palette knife.

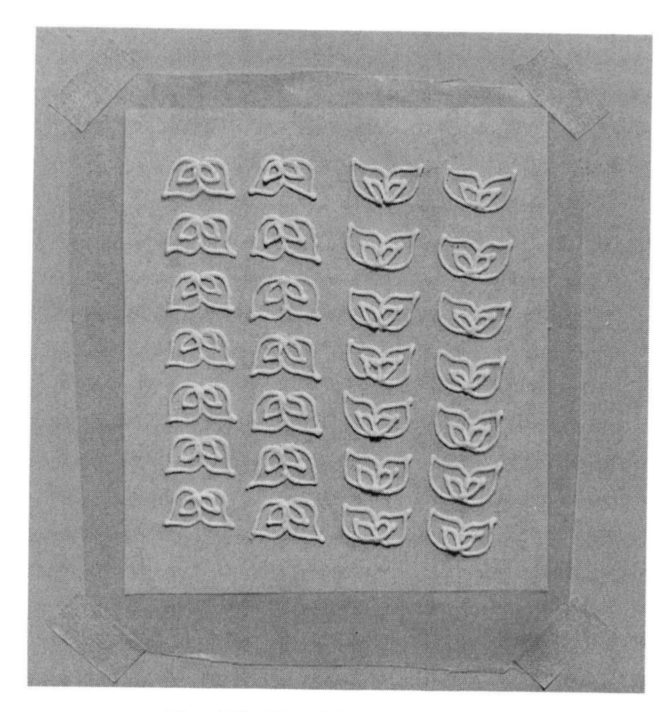

Fig. 52. Piped butterfly wings.

To Assemble Butterfly

Assemble the wings directly on to the cake by piping a bulb of icing with a No. 1 tube and placing the wings into it. Bodies are not required for very small butterflies, but if desired can be added with a No. 1 tube (No. 2 tube for large butterflies) after they have been assembled.

Piped butterflies are very light and will not require any support whilst being assembled.

Working drawings for petals and butterflies are given in Fig. 53.

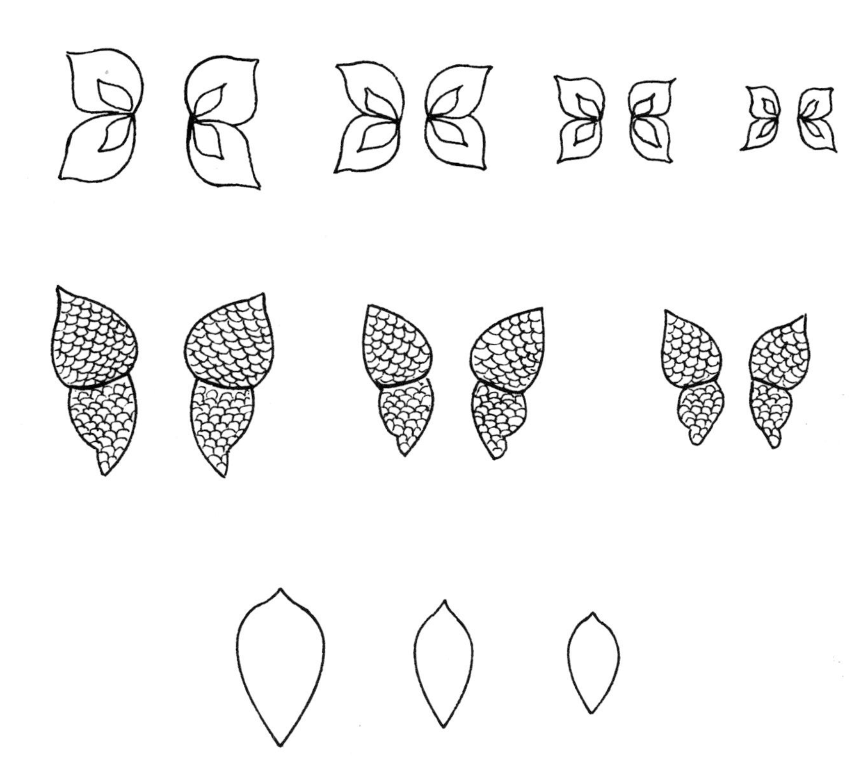

Fig. 53. Petals and butterflies — working drawing.

Corner Pieces

It is usual to decorate the corners of square cakes, although this is not absolutely necessary if the coating is first class.

One form of decoration is to pipe a small shell down the corners using a No. 2 or No. 3 tube. Another is to attach corner ribbons as illustrated in *The Art of Royal Icing*.

The designs shown in this chapter are what I term as 'curly corner pieces' and are piped using a No. 2 and a No. 1 tube.

Advance Preparation

Unlike other runouts these pieces can not be piped until the depth of the coated cake is ascertained.

1. Measure the depth of the cake and reduce this by a small amount to allow for the pieces to be placed underneath the top border.
2. If the pieces are to be placed directly into the flooding on the board, the measurement should be taken *before* the bottom board is flooded.
3. If the pieces are to be placed on top of the flooded board, measurement takes place after the board has been flooded and is *completely dry*.
4. Several designs are shown in Fig. 57 but should they not be the size required, it is a simple matter to alter a drawing to suit your needs.

(a) Trace the required drawing from the book. Measure the depth of the cake and draw a line the correct size on a piece of paper.
(b) Trace the corner piece on to the line and adjust the drawing by adding another scroll or by making part of the design shorter (Fig. 54).
(c) When satisfied with the result, go over the design with a pen and trace as many as required.

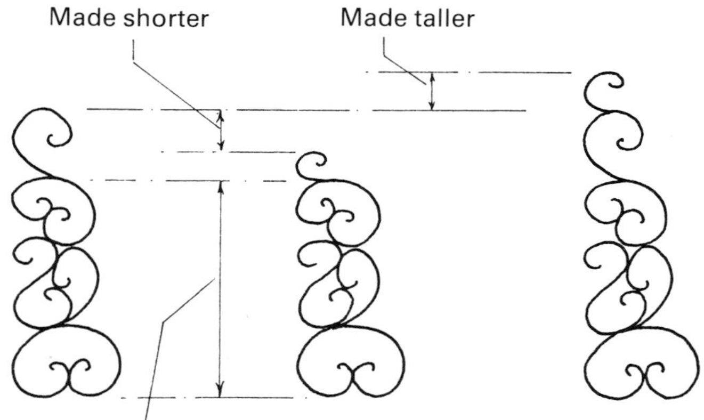

Fig. 54. Adjusting height of corner pieces.

5. If a design is to be used for a cake with tiers of varying depths, make sure the designs are similar in shape even though different in size.

Points to Remember

1. Make more pieces than required to allow for breakages.
2. Take care to see that every section is joined, otherwise the design will not hold together when removed from the waxed paper.

Procedure for Piping Corner Pieces

1. Place tracings on a board, cover with a piece of waxed paper, shiny side up, and stick the corners down with masking tape.
2. Outline with a No. 2 tube using *touch, lift* and *place* method of piping.
3. Allow to dry a little and overpipe, still using a No. 2 tube.
4. When crusted, pipe a small shell over the plain linework using a No. 1 tube.
5. Leave to dry for several hours or approximately one hour if under a lamp.
6. Slide an artist's palette knife *very gently* under the piping to separate the scroll from the paper and carefully turn over.
7. Complete by overpiping a small shell with a No. 1 tube.

To Assemble Corner Pieces

Small bulbs of icing on the edge of the corners (where they touch the cake) and underneath (where they meet the cake board) is all that is necessary to attach them to the cake. A paint brush is used to remove any surplus icing before the icing is allowed to dry.

Corner pieces are sometimes placed into a shell border that has been piped down the corners of

Fig. 55. Corner pieces in flooded cake board.

the cake. If this method is used, the shell should be piped with a No. 2 tube and the corner piece placed into it before the shell has crusted.

If the corner pieces are intended to be placed into the flooded cake board, this must be done *before* the icing has been allowed to dry. A disadvantage of this method is that if the corner piece is broken after it has been assembled, it is almost impossible to repair satisfactorily. Figure 55 shows corner pieces placed into a flooded cake board.

Corner pieces placed on a dry board are assembled *after* all other decoration has been completed (Fig. 40).

Corner Pieces With Flooding (*Fig. 56*)

Two of the designs (f and g) in Fig. 57 look very attractive if they are partially flooded.

(f) Outline twice with No. 2 tube and flood the heart with softened sugar before overpiping the entire outline with a No. 1 shell. Leave to dry for *several hours*. Turn over, flood again and pipe the shell edging with No. 1 tube.

(g) Pipe with No. 2 tube twice and No. 1 shell

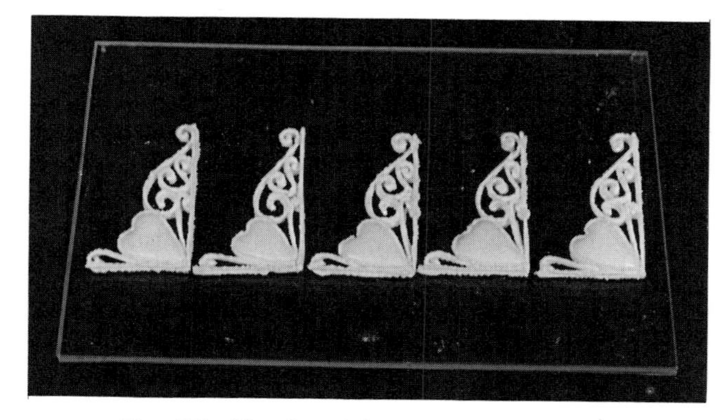

Fig. 56. Piped and flooded corner pieces.

once, ignoring the flowers and leaves. Flood the leaves and alternate petals of the flowers (the petals cannot be flooded in one movement otherwise they will lose their shape.) Complete flower when adjoining petals have crusted and edge the petals and leaves with a No. 1 tube, when dry. Turn over and pipe No. 1 shell over the scroll work. Flood leaves and flowers as before, and finally, outline them with a No. 1 tube.

Working drawings for corner pieces are given in Fig. 57.

Fig. 57. Corner pieces — working drawing.

CHAPTER 10

Outlined Runouts

Outlined runouts fit in to one of two categories.

1. Those where outlining is *preferred* (lettering, numerals and small motifs where a well defined edge is required).
2. Those where outlining is *essential* (certain figures, all runout borders and plaques).

The figure drawings to be found in Fig. 63 appear in this chapter because they look more realistic with a dark outline, in spite of the fact that they are not realistic figures! Rupert Bear must be outlined and therefore other motifs that appear with him (drum, football, kite for instance) need to be outlined also. If these motifs were to be used with figures that did not have outlines, their own outlines could be dispensed with.

Fig. 58. Golden wedding cake — 20 cm (8 in.) cake on 30 cm (12 in.) board.

Lettering, Numerals, Motifs etc.

Runout numerals and lettering have many uses in cake decorating. It is much easier to achieve a good result by running out an inscription rather than directly piping it on to the cake. Since runout lettering takes up more space than direct piping however, it is essential that it is kept to the minimum and that the size of the lettering is suitable for the size of the cake.

Monograms are combined letters and practice is required in placing them together. When a monogram is to be placed on the side of a cake, the depth of the cake must be taken into consideration at the design stage.

Lettering dried flat will fit the side of any square cake, but in order to fit a round cake, it must be kept quite small. If larger letters are required for the side of a round cake they must be dried on a curve similar to that of the side of the cake.

A few letters are impossible to combine well. If

Fig. 59. Twenty first birthday cake — 18 cm (7 in.) cake on 28 cm (11 in.) board.

this should be the case, I suggest that separate initials are runout; then placed on the cake or plaque and linked together with a heart or bow.

Runout lettering and numerals are suitable for mounting on to plaques or may be placed directly on the cake and stuck with a little icing. The inscription on the Christmas cake (Plate 1) combines both of these methods. The golden wedding cake (Fig. 58) shows how numerals can form part of the design; the numerals on the twenty-first cake (Fig. 59) are its main feature.

Points to Remember

1. All outlines are made with icing of piping consistency in a small bag fitted with a No. 1 tube.
2. Where outline and flooding are the same colour, the outline should not be visible. This is achieved by piping one item at a time and flooding *immediately*.
3. A silver or gold edge to motif, numeral or letter is achieved by piping the outline and allowing it to crust before painting silver or gold and allowing the paint to dry before flooding. Any paint on the waxed paper will be left behind when the runout is removed. This method is very attractive with horseshoes, stars and icicles.
4. An edging of contrasting colour is obtained by outlining with the desired colour and allowing it to crust before flooding with icing of a different colour.
5. Some lettering, numerals and small figures, such as a vase, may be flooded in one movement. Some however, will require parts to be carried out separately and allowed to crust in order that they be easily recognisable.
6. Very little icing is used when flooding small areas therefore small bags are used.

Procedure for Piping Lettering, Numerals etc.

1. Remove icing required for outlining and dilute a small amount of the remainder with egg white or albumen until a soft, but not runny, consistency is reached. Place in a small bag.
2. When ready to commence work, check the consistency of the runout icing. To do this, cut a small hole in the bag, not larger than a No. 2 tube and press out a little icing on the work surface or saucer. Use a paint brush to move the icing, which should not flow but be soft enough to lose the brush marks.
3. Pipe an outline with a No. 1 tube and if the outline is not required to be visible, flood immediately. The flooding should be brushed on (but not over) the outline with the paint brush. If outlines of a different colour to the flooding are required; pipe all and allow to crust before flooding, taking care not to cover the outline (Fig. 60).

Fig. 60. Monograms outlined prior to painting and flooding.

4. Place under a lamp before transferring to a warm place to dry.

Working drawings for outlined runouts are given in Fig. 61.

Decoration

Decoration may be added but this is not carried out until after the runout is completely dry. Small flowers or leaves may be piped or a fine edging added to a plain monogram as shown in Fig. 74. All piping is carried out with No. 0 tubes.

Outlined Figures

On the whole I do not like an outline to show on a runout figure, but there are exceptions to this.

(text continued p. 71)

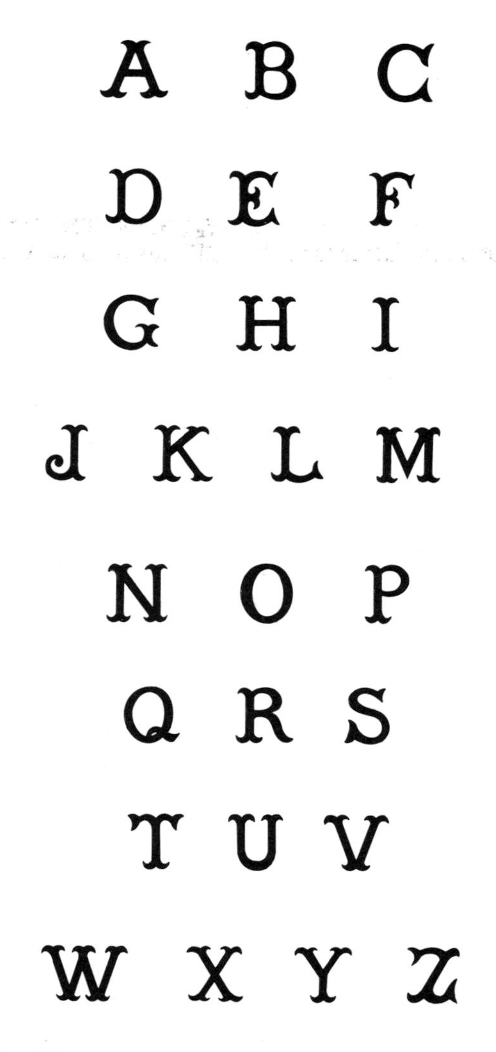

Fig. 61. Working drawings for outlined runouts.

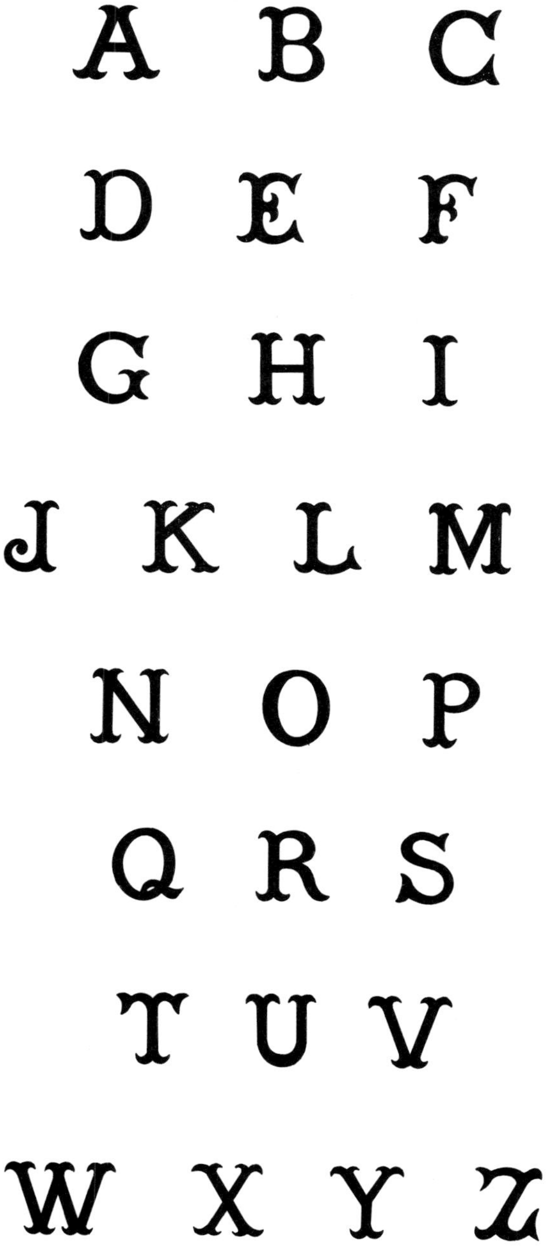

A B C

D E F

G H I

J K L M

N O P

Q R S

T U V

W X Y Z

Fig. 61 — *contd.*

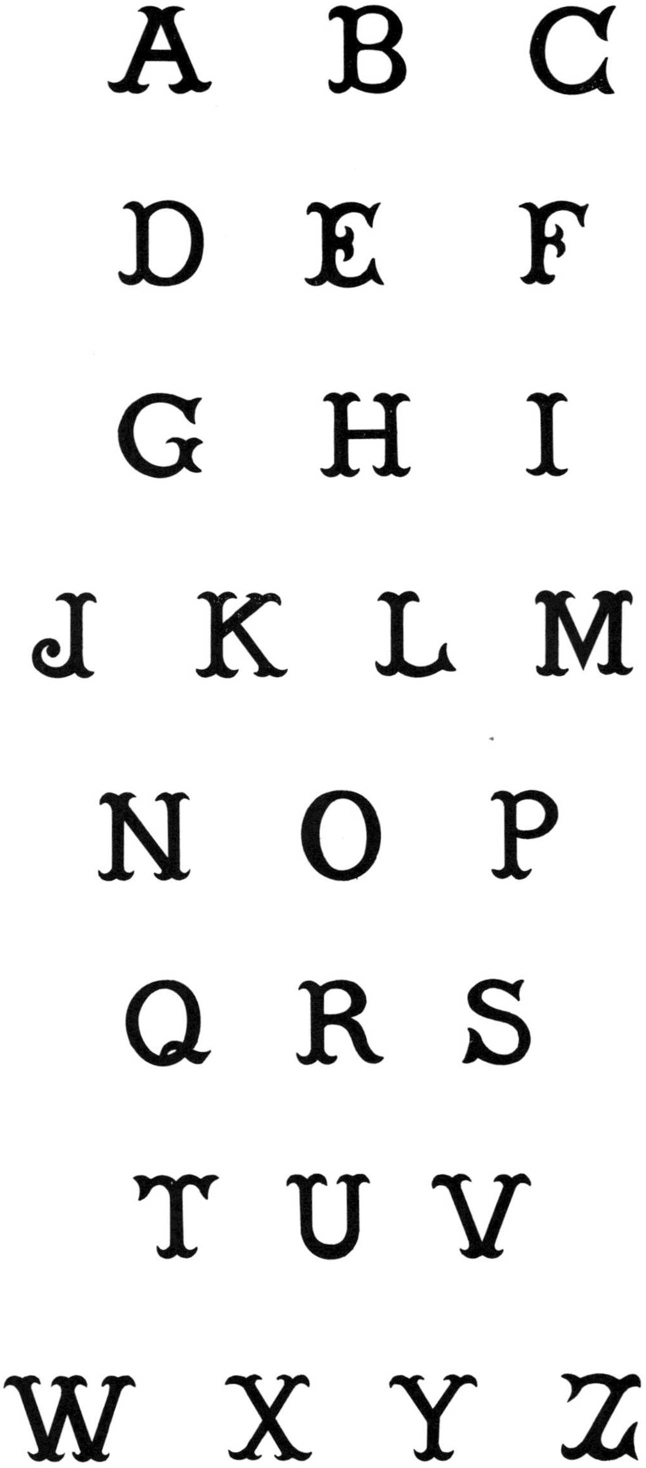

A B C

D E F

G H I

J K L M

N O P

Q R S

T U V

W X Y Z

Fig. 61 — *contd.*

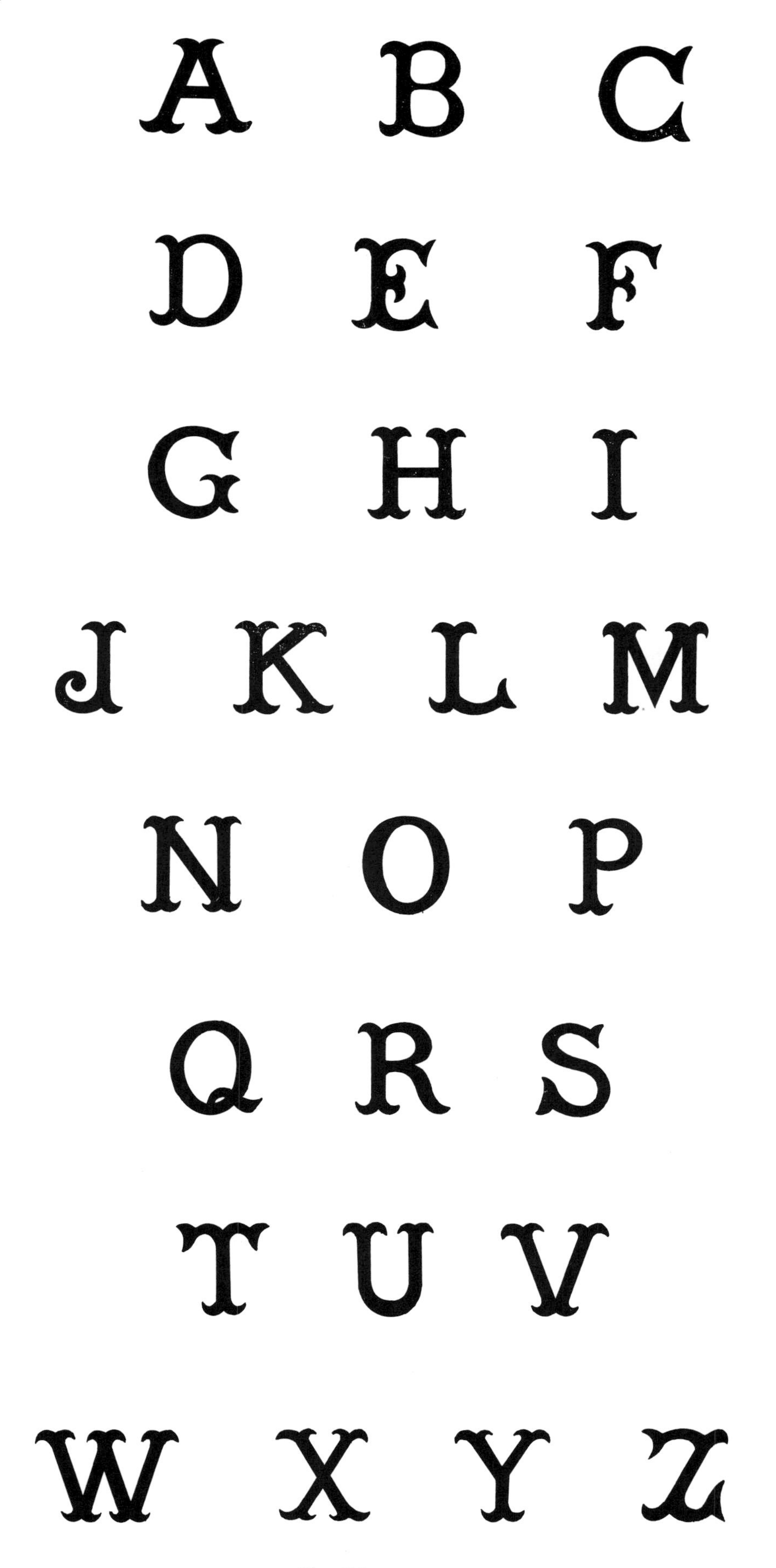

Fig. 61 — *contd.*

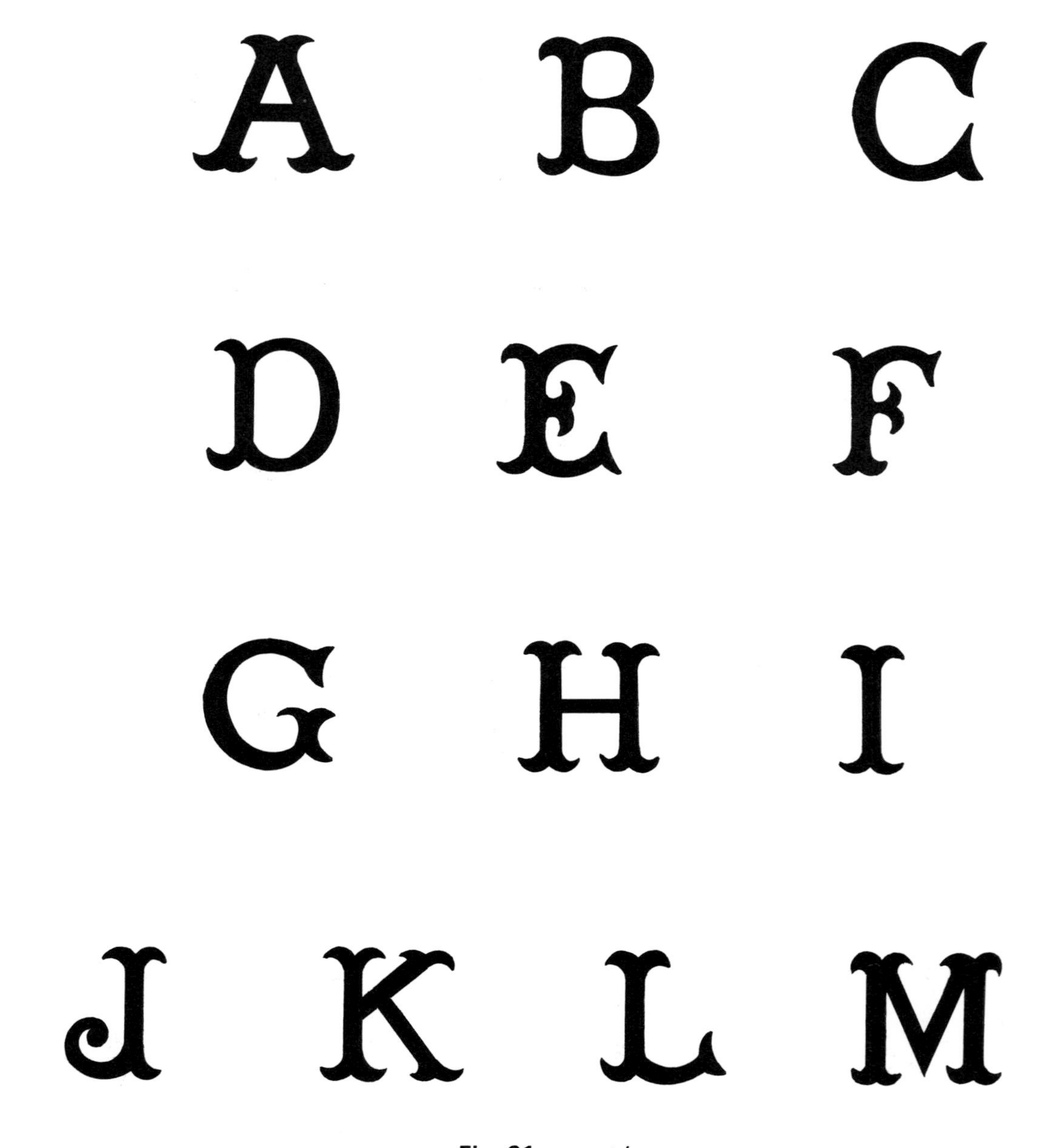

Fig. 61 — *contd.*

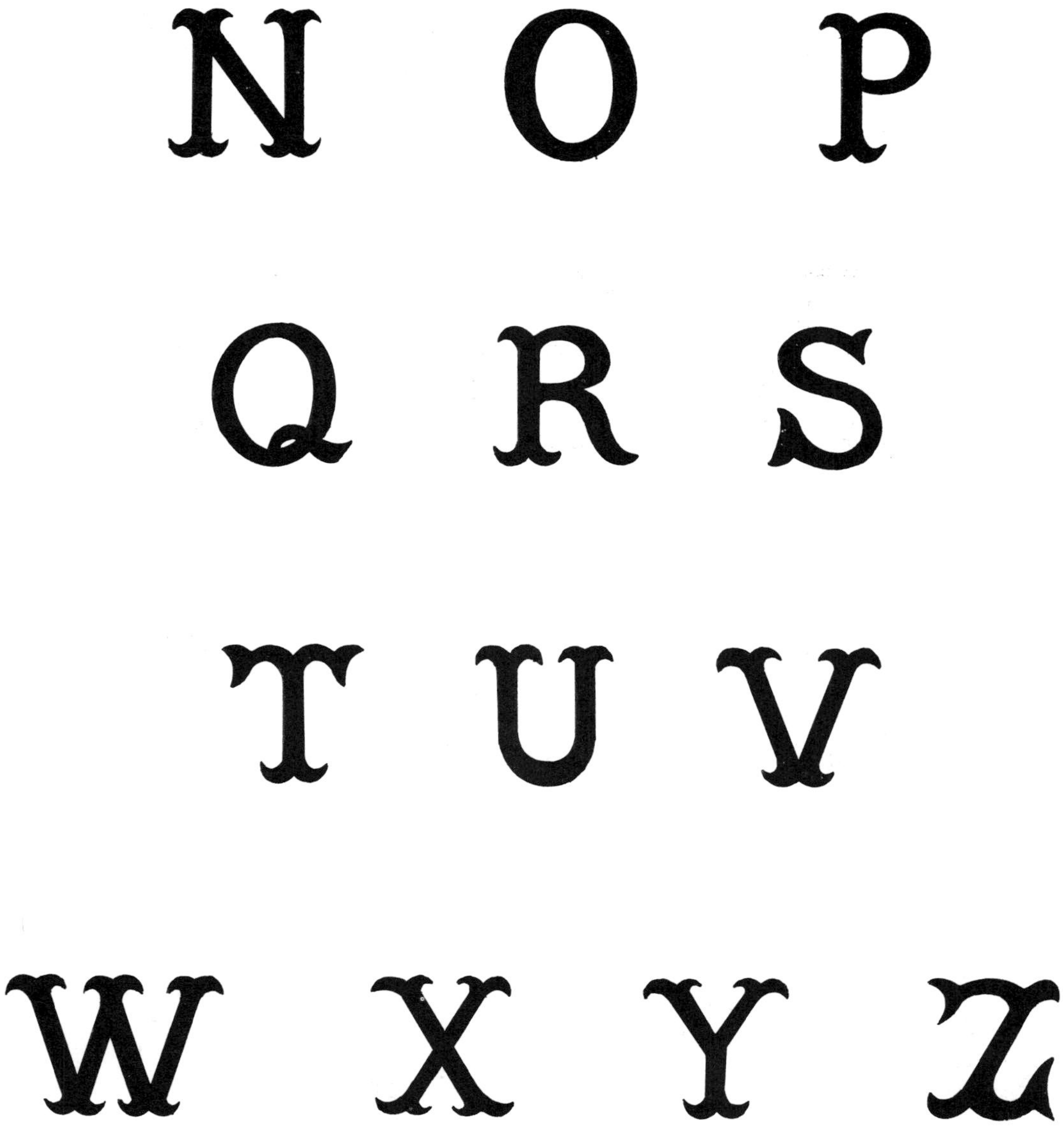

N O P

Q R S

T U V

W X Y Z

Fig. 61 — *contd.*

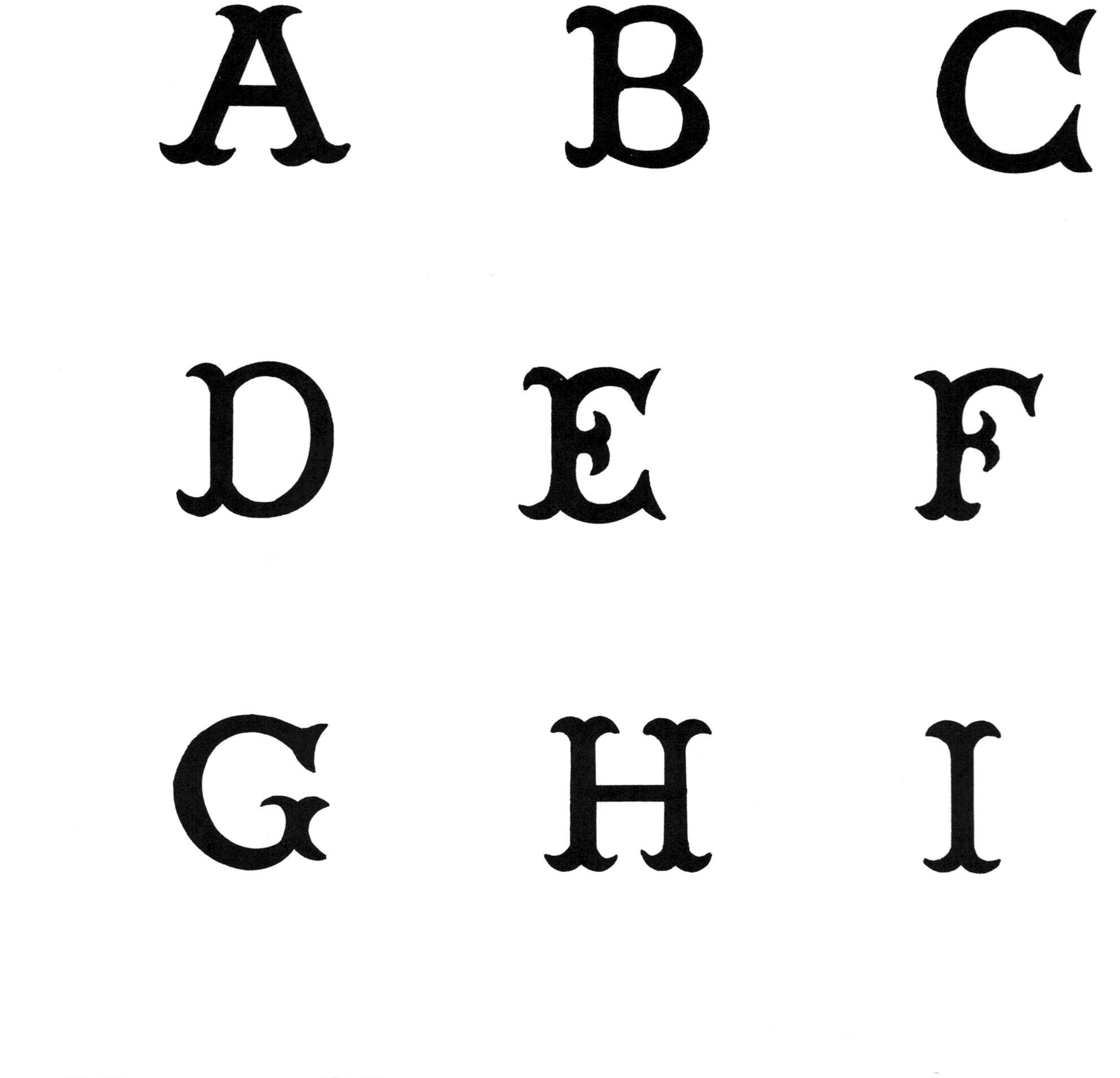

A B C

D E F

G H I

J K L M

Fig. 61 — *contd.*

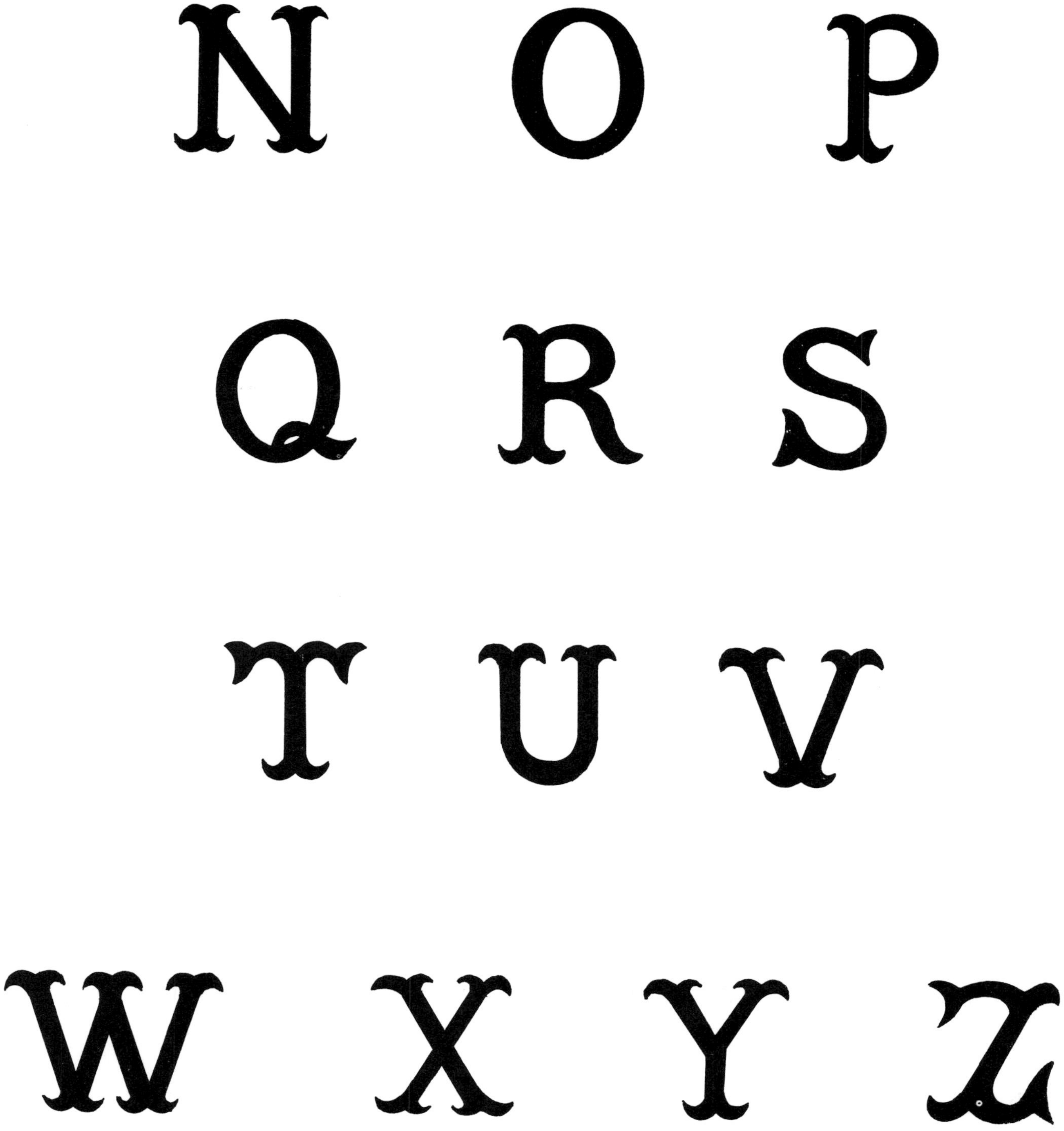

N O P

Q R S

T U V

W X Y Z

Fig. 61 — *contd.*

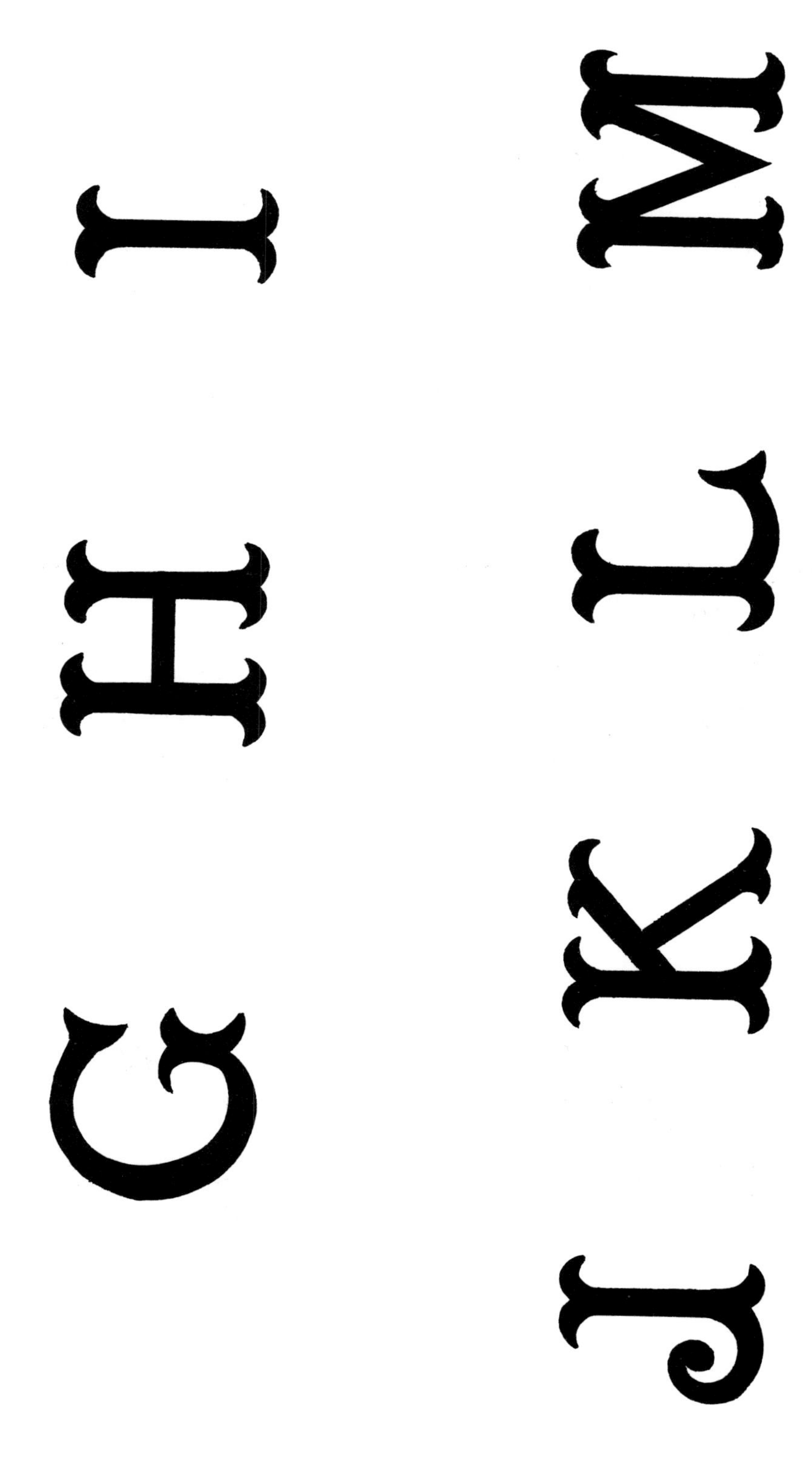

Fig. 61 — *contd.*

N O P

Q R S

Fig. 61 — *contd.*

Fig. 61 — *contd.*

N O P

Q R S

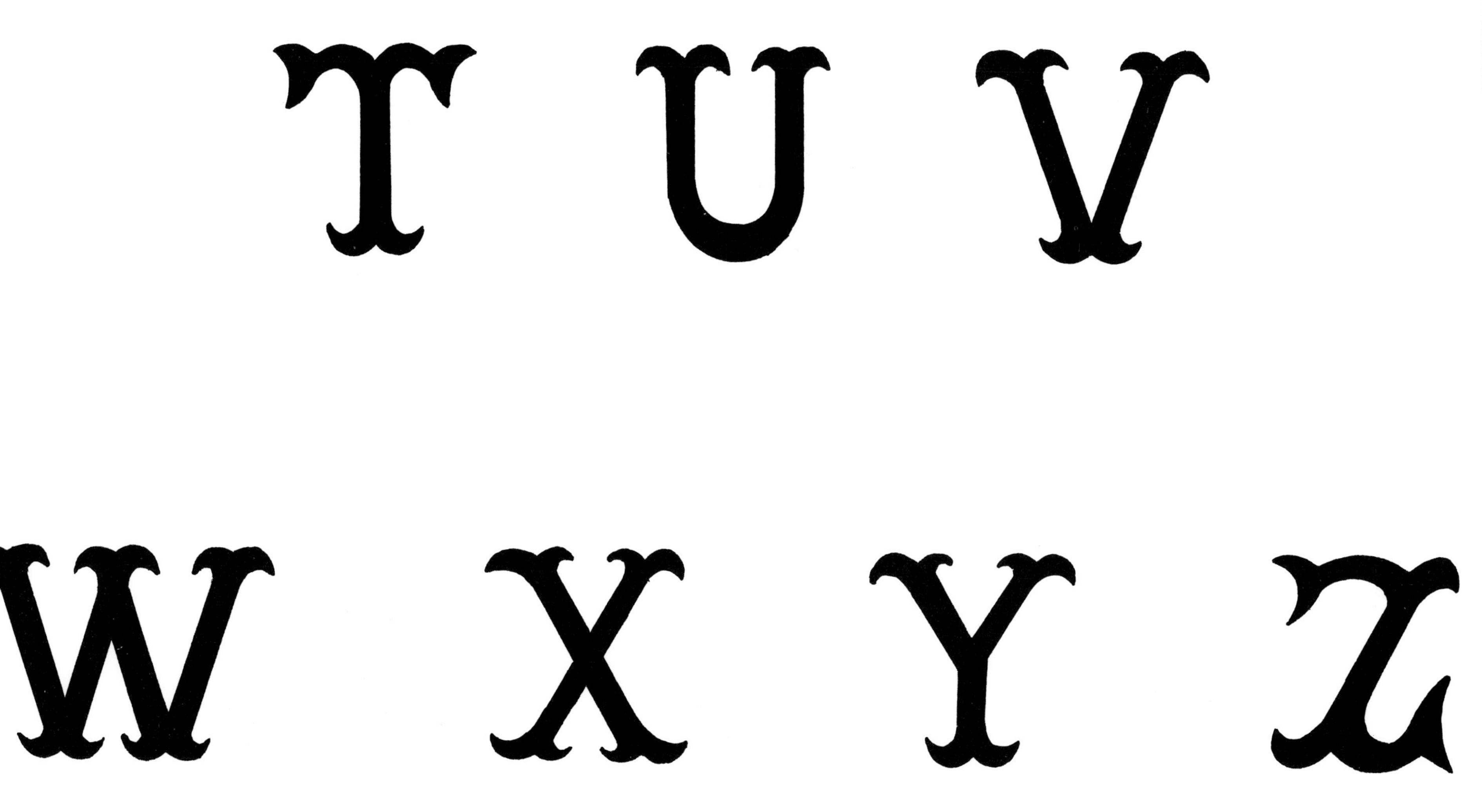

Fig. 61 — *contd.*

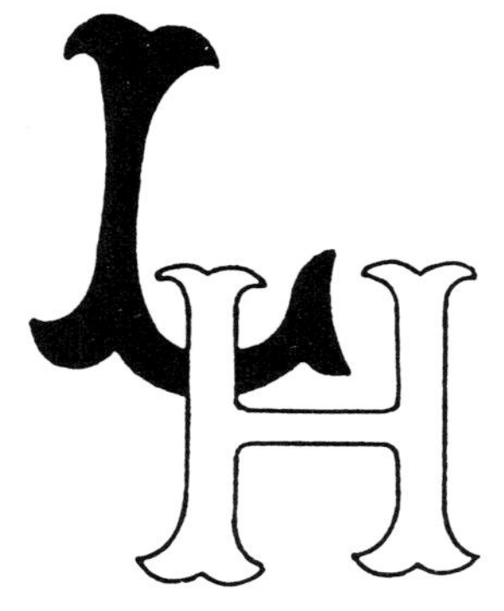
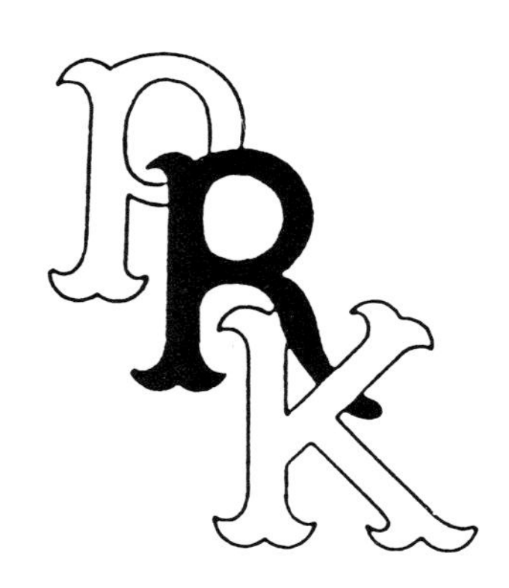

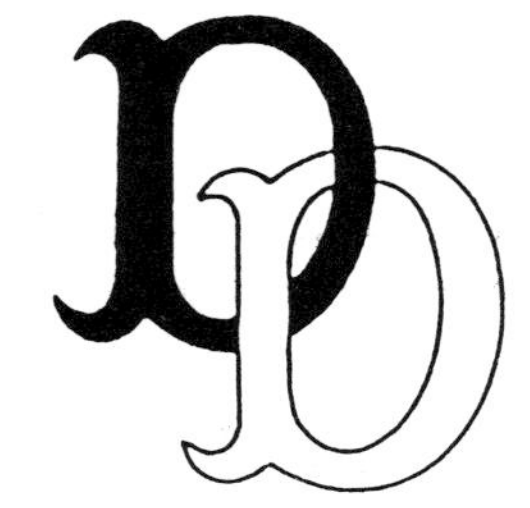

Fig. 61 — *contd.*

Fig. 61 — *contd.*

12345
67890

12345
67890

12345
67890

Fig. 61 — *contd.*

12345
67890

12345
67890

12345
67890

12345
67890

12345
67890

Fig. 61 — *contd.*

Fig. 61 — *contd.*

Fig. 61 — *contd.*

Fig. 61 — *contd.*

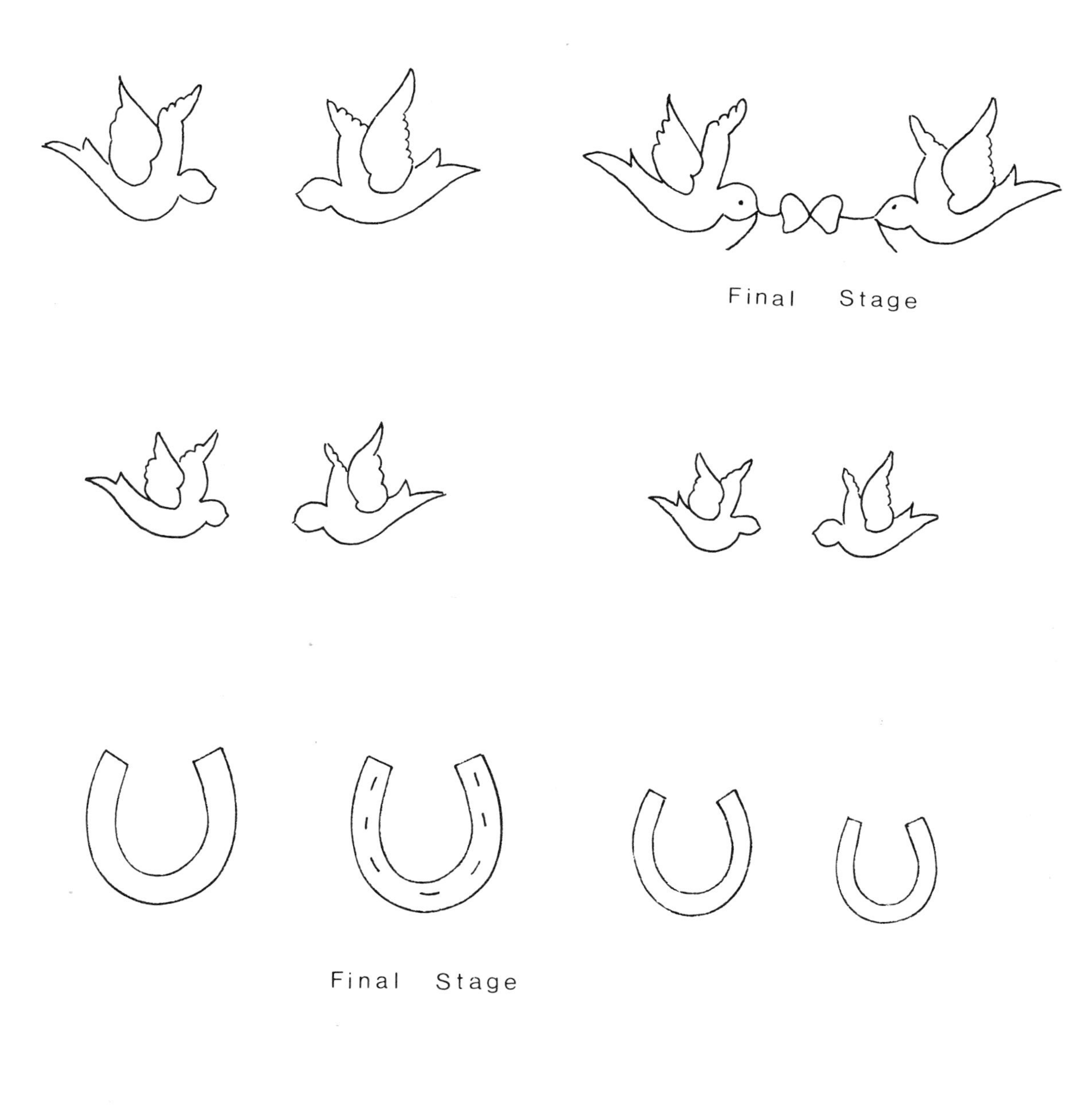

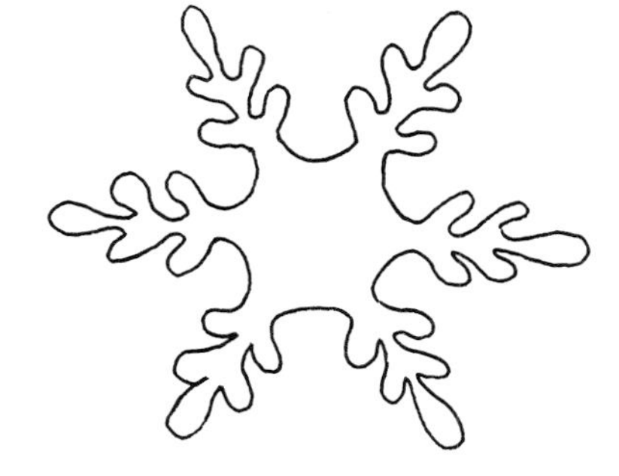

Fig. 61 — *contd.*

Fig. 61 — *contd.*

FINAL STAGE

Fig. 61 — *contd.*

Fig. 61 — *contd.*

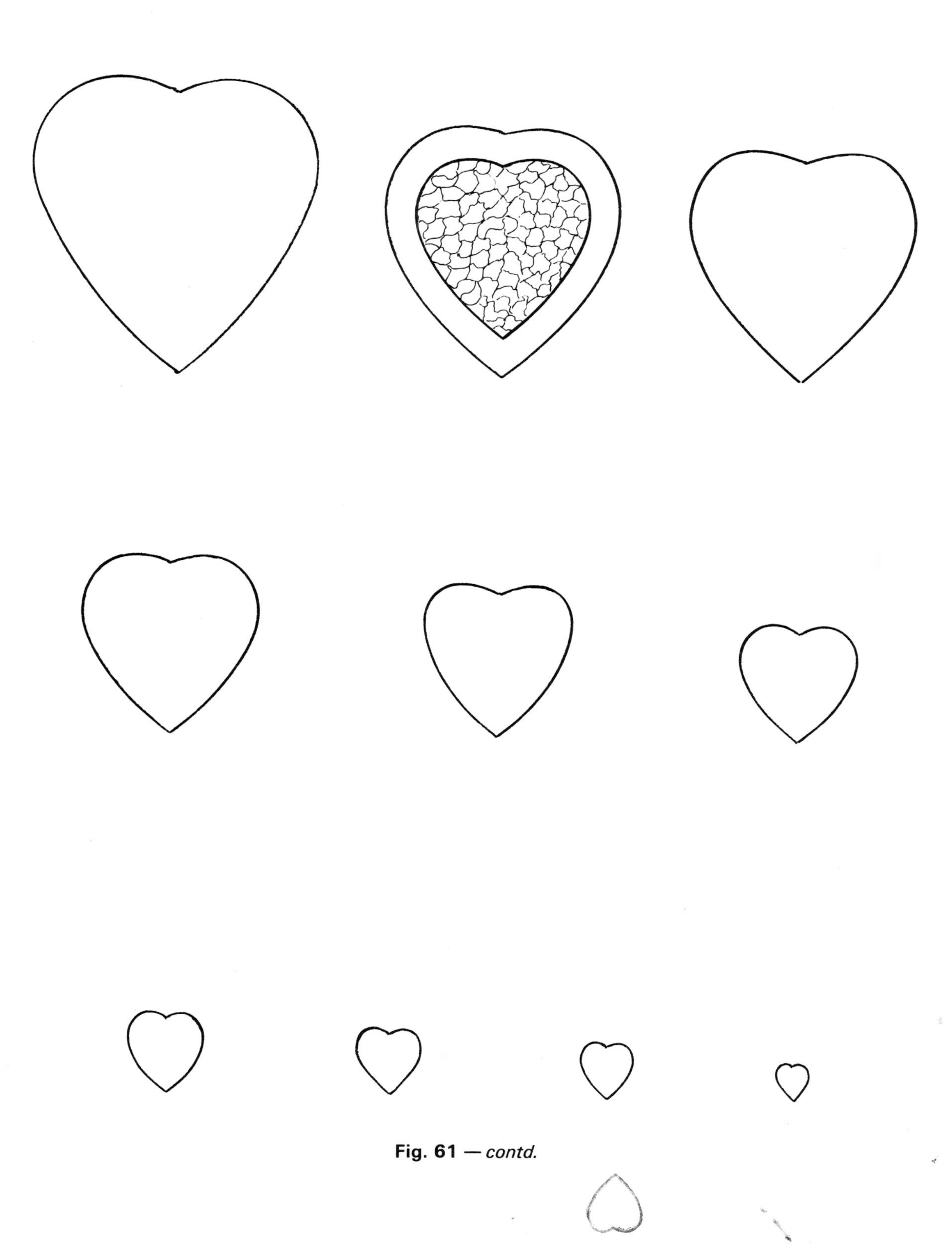

Fig. 61 — *contd.*

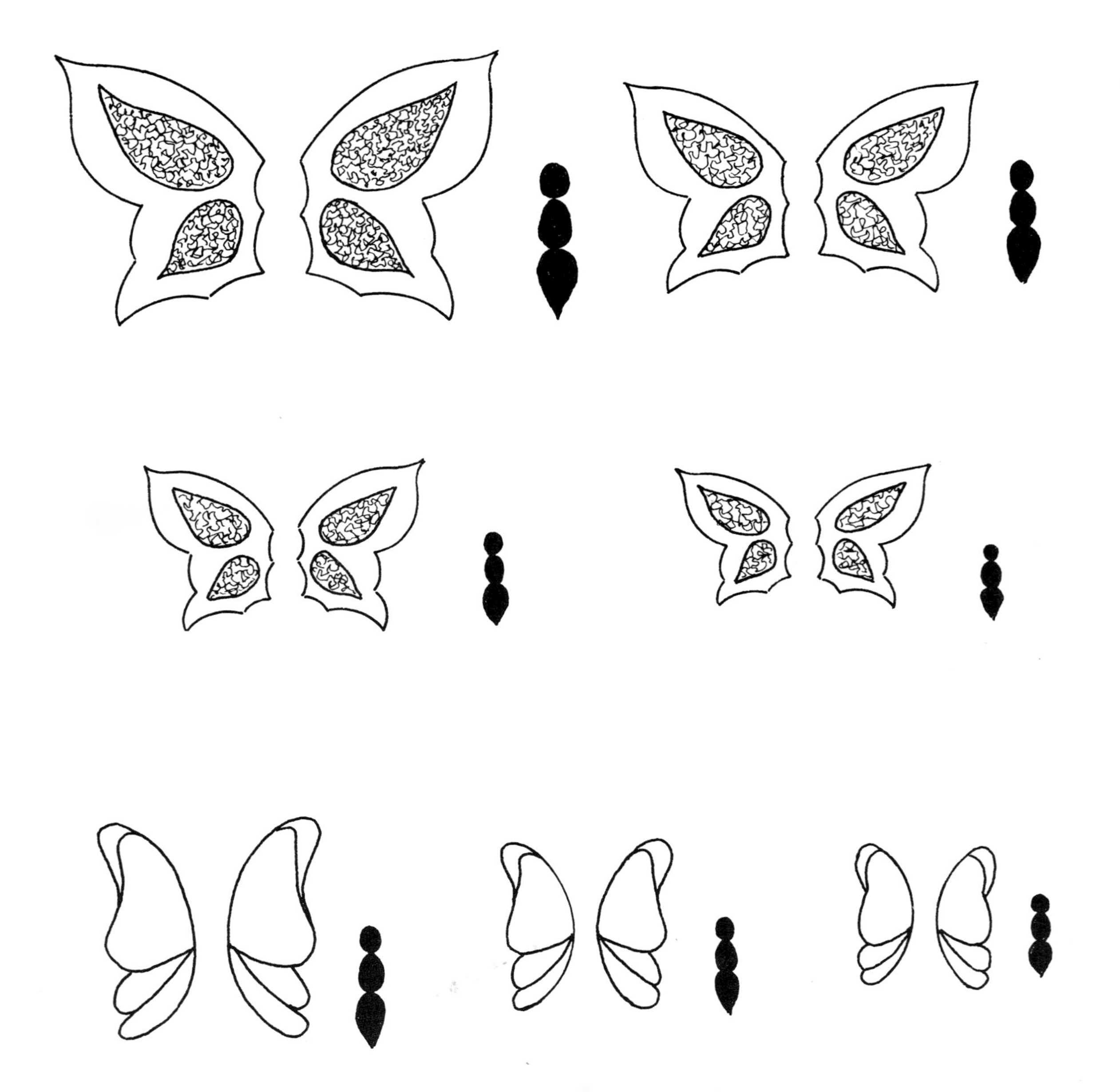

Fig. 61 — *contd.*

Outlined figures for use on childrens cakes are very agreeable and Rupert Bear is a perfect subject for this method of piping.

Not only are Rupert and his friends straightforward to pipe, but they are firm favourites with children of all ages. The simple addition of a kite, balloon or banner depicting the child's name or age is all that is necessary to complete the design. I am sure that Rupert Bear was designed with royal icing in mind!

In addition to these figures, others suitable for this method of piping are shown in Fig. 63.

Points to Remember

1. Outlines are made with a No. 1 tube with icing of piping consistency.
2. Outlined figures are flat and therefore the icing used for flooding is thinner than that used for other methods.
3. Dark colours (brown or black) are used when outlining, and bright colours (where appropriate) for flooding. Coating and border work must be kept pale.

Procedure for Piping Outlined Figures

1. Outline all figures using the *touch, lift* and *place* method and leave to crust whilst preparing the icing required for flooding (Fig. 62).

Fig. 62. Piped outline of Rupert Bear. Rupert Bear is reproduced by kind permission of the *Daily Express.*

2. Colour the icing that is required for flooding and reduce to a consistency that will flow gently without requiring the aid of a paint brush.
3. Place in small bags. If a large area is to be flooded it is preferable to use two small rather than one large bag.
4. Cut a small hole in the bag, not larger than a No. 2 tube and before commencing, check that the consistency is correct.
5. Flood as many parts as can be done without coming in contact with another section.
6. Care should be taken that the icing does not overflow the linework. A paint brush will be required to gently ease the icing into the corners.
7. Allow to crust under a lamp whilst flooding the next figure.

Features must not be painted until the runouts are completely dry. Additional decoration to outlined figures (on clothes etc.) must be *painted* and not piped. This should be carried out with a fine paint brush (kept as dry as possible) using edible colourings or with an edible pen (Plate 2).

These figures are very fragile and should be handled with care. Leave on boards until required and remove with an artist's palette knife or a piece of cotton thread, held taut between both hands and passed between the underside of the figure and the waxed paper.

Working drawings for outlined figures are given in Fig. 63.

Runout Borders

There are two types of runout borders.

(a) A continuous band or 'collar'.
(b) Runout pieces.

Both types are suitable for round or square cakes, but it is easier to make separate pieces and assemble them on the cake than to make a collar and place it in one piece on the cake.

Details of how to design runout borders are shown in *The Art of Royal Icing*.

Points to Remember

1. Runout borders are outlined with a No. 1 tube, using the *touch, lift* and *place* method.
2. Great care must be taken when releasing a large collar from the waxed paper as they are easily broken. For this reason, do not use complete collars for cakes of 25 cm (10 in.) and upwards.

(text continued p. 83)

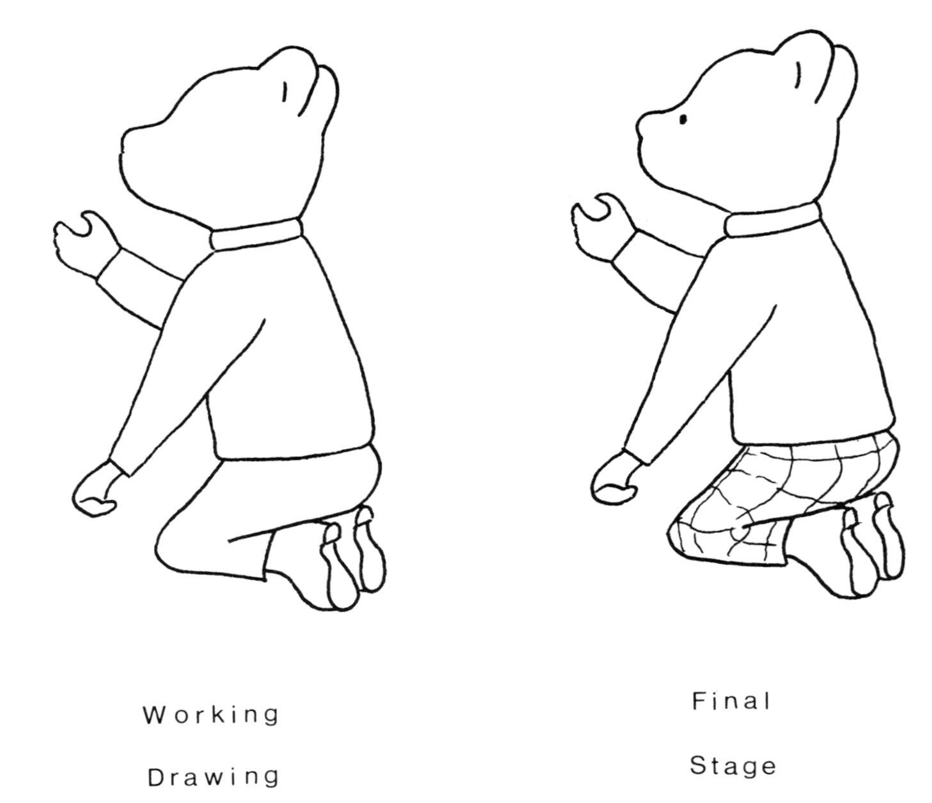

Fig. 63. Working drawings for outlined figures.

Working Drawing

Final Stage

Fig. 63 — *contd.*

Working
Drawing

Final
Stage

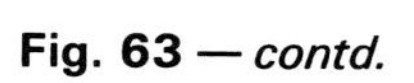

Fig. 63 — *contd.*

Working Drawing

Final Stage

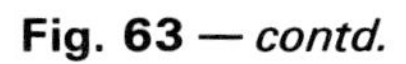

Fig. 63 — *contd.*

Fig. 63 — *contd.*

Fig. 63 — *contd.*

Fig. 63 — *contd.*

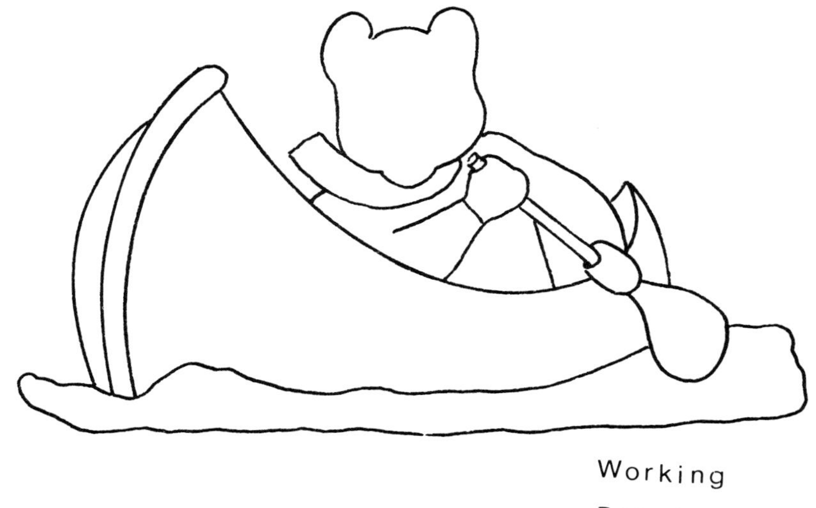

Working Drawing

Intermediate Stage

Final Stage

Fig. 63 — *contd.*

Final Stage

Fig. 63 — *contd.*

Working Drawing

Final Stage

WORKING DRAWINGS

Fig. 63 — *contd.*

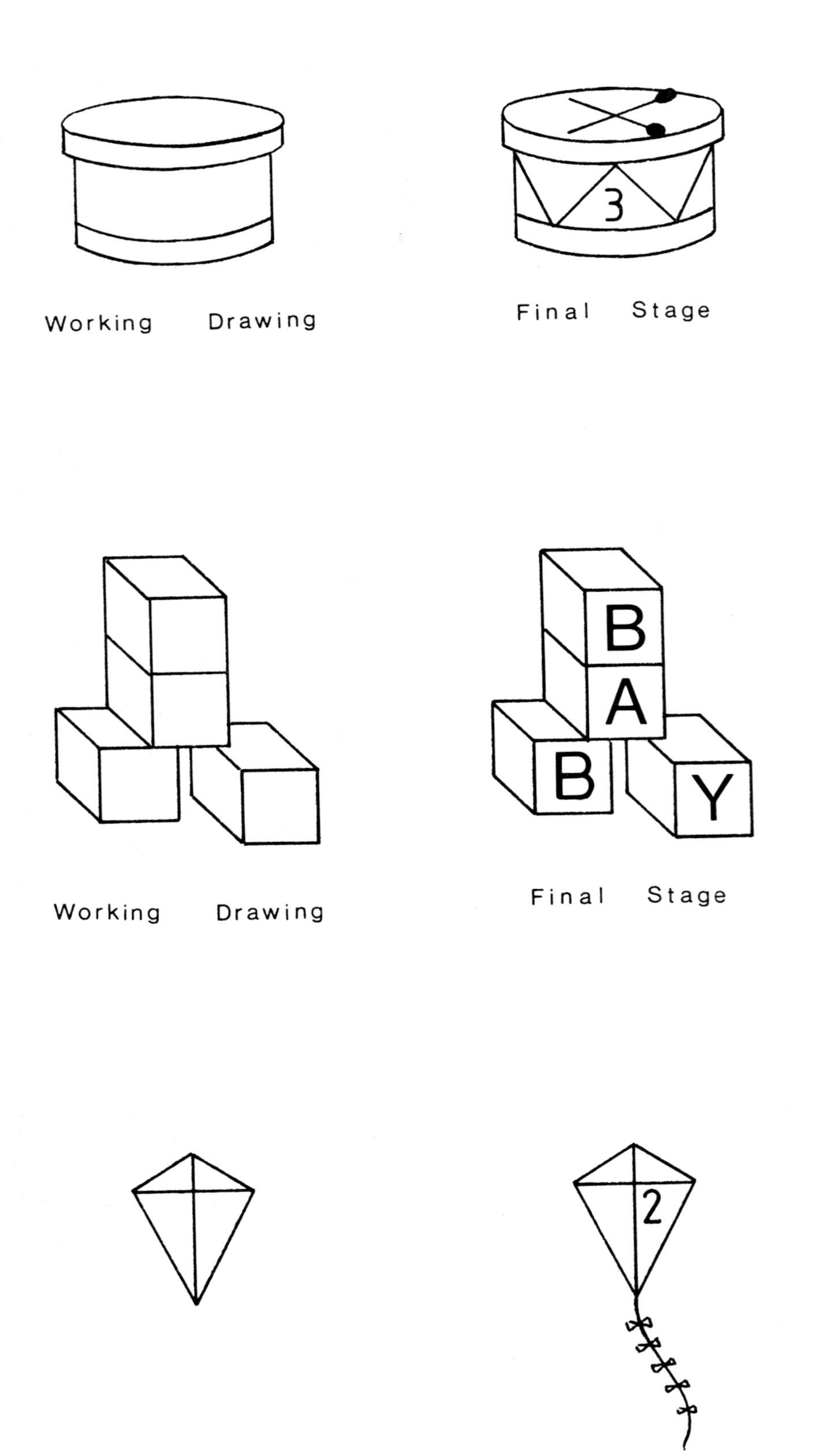
3
Working Drawing
Final Stage
B
A
B
Y
Working Drawing
Final Stage
2
Working Drawing
Final Stage

Fig. 63 — *contd.*

3. Clearly indicate on the tracing the size of cake for which the border is intended.
4. If a collar is made, the waxed paper must be cut in the middle to allow for expansion.
5. Icing should be mixed to piping consistency; diluted with albumen and left for several hours to allow air bubbles to disperse.
6. A border is flooded before the outline is allowed to crust.
7. If *filigree* is to be included in the border, the shape of the inner design is piped first with a No. 1 tube. Filigree is then piped with a No. 0 tube, making sure that it touches the linework at frequent intervals. The outer linework is piped and the border flooded immediately.
8. Where *flowers, leaves, numerals etc.* are to be *included with filigree* they are carried out before the inner linework. Additional piping and painting (stamens, shading etc.) is left until the border is finished and dried.

There are two methods of piping flowers and leaves.

(a) *Flooding* the shapes with soft icing in the appropriate colours. The hole in the bag should be about the size of a No. 1 tube. Alternate petals of flowers are flooded and allowed to crust before the adjoining petals are flooded. This ensures that each petal retains its shape.
(b) *Piping* on to the designs with No. 1 tubes; the consistency being the same as for the linework (Figs 64 and 65). The tube is held near the surface and the petals piped using a side-to-side movement of the wrist.
Leaves are formed by pressing the tube to form a bulb and pulling back to produce a 'tail'. Veins are made with a paint brush before the icing is allowed to crust.

Whichever method is used, flowers and leaves of bright colours should be allowed to crust before piping filigree thus avoiding the colour bleeding. In addition to touching the linework, the filigree must also make contact with the flowers and leaves otherwise they will collapse when released from the waxed paper.

Flowers painted silver or gold are piped in the same colour as the border and allowed to dry before being painted. This takes place before the addition of filigree, thus avoiding the paint touching the icing.

Numerals are outlined with a No. 1 tube and flooded before filigree is piped. If they are to be painted they must be completely dry before this takes place (Fig. 58).

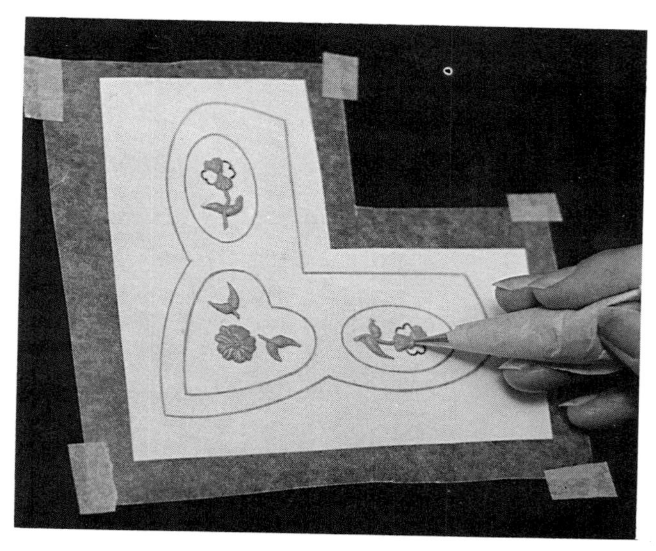

Fig. 64. Piped flowers in border piece.

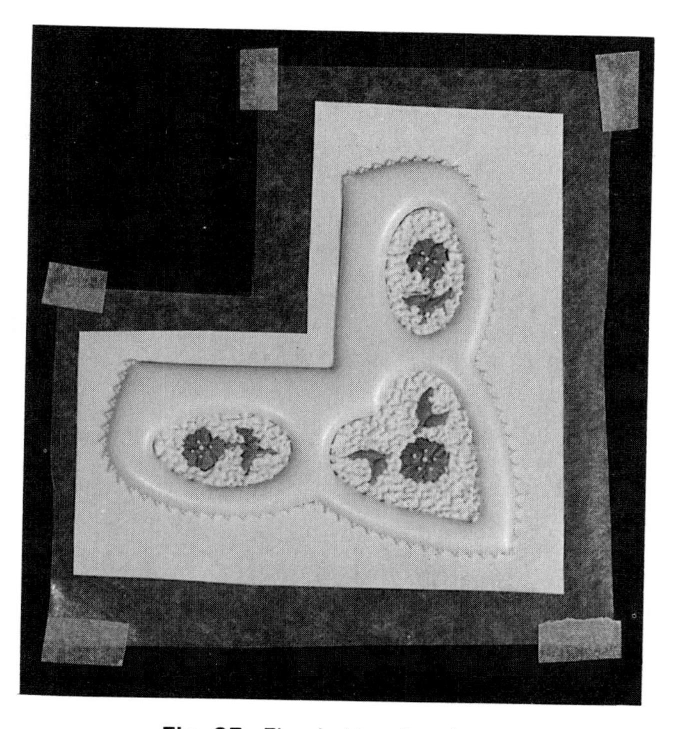

Fig. 65. Flooded border piece.

9. In a border where the *decoration* is *unsupported by filigree* the decoration is carried out before the border is flooded. Care must be taken when piping the inner linework to ensure that it comes in to contact with the decoration as this is its only form of support (Plate 3).
10. *Netting* and other forms of linework are *piped first* before the inner shape is outlined. It is taken slightly beyond the edge of the design. The inner linework is piped over the 'tails' of the netting thus ensuring a neat finish to the border (Fig. 66).

Fig. 66. Decoration suitable for inside runout borders.

11. Always make more pieces than required, and two collars instead of one. Spare borders can be used another time.

12. It is possible to use a drawing for a corner piece intended for a smaller cake on a larger cake. It is not possible, however, to use a drawing intended for a larger cake on a smaller one. Generally, complete collars must be used only for cakes of the size indicated on the drawing.

Procedure for Piping Runout Borders

1. Pipe flowers, numerals, netting, filigree etc. if any. This can, if you wish, be carried out several hours (or days!) in advance of flooding.

2. Have ready several large bags half filled with softened icing. This will be required to be slightly stiffer than for outlined figures as a cushioned, rather than flat, runout is required.

3. Before flooding, cut a small hole in the end of the bag about the size of a No. 3 tube and test the icing for consistency. It should be the same as that used for lettering, numerals etc. – soft but not runny.

4. Outline the collar or border piece with a No. 1 tube, making sure there are no gaps for the icing to escape.

5. Flood thickly and evenly using a fine paint brush to assist the icing into the corners and to cover the outline. Only 5 cm (2 in.) should be flooded at a time and made smooth before adding another portion. If a continuous border is being executed, alternate portions should be flooded in the same way as for the cake board (see page 32).

6. Leave under a lamp, for as long as possible, before putting in a warm place to dry.

7. The border must be thoroughly dry before an edging is piped. The outer edge is loosened a

little from the waxed paper before being edged with a No. 0 tube. It should remain on the waxed paper; still attached to the board, until it is required.

To Assemble Borders

1. Runout borders are stuck on to the cake with a little icing. When placing a collar on a cake, I pipe a line on to the edge of the cake with a No. 2 tube and place the collar gently on the top. For runout pieces, I find it easier to turn the runout over and place a little icing on the back before securing it to the cake. Surplus icing must be removed with a paint brush before it is allowed to dry.

2. Some form of linework is required *underneath* the runout, where it sits on the cake. A shell is piped, very carefully, *after* the border has been assembled. Linework of scallops or loops are piped *before* the border is assembled (Fig. 67).

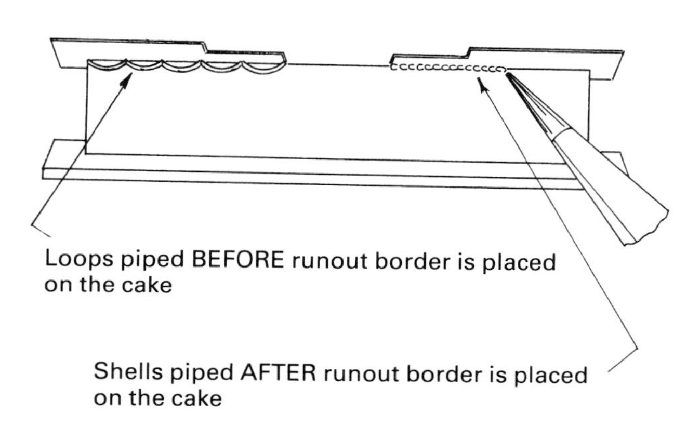

Fig. 67 Loops and shells.

3. Removing the waxed paper from a small runout is a simple matter, but not so with large collars. If a collar does not have an edging of bulbs, it can be removed by pulling a taut piece of thread between the collar and the waxed paper. This method is unsuitable for runouts with bulbs and the easiest way is to keep the border on the board and place it on to a turntable. Remove the tape and move the collar gently until one part is over the edge of the board. Peel the paper from the runout using a downward movement. Turn the turntable a little and repeat until the whole runout is free.

4. Linework is often required inside the border and I like to follow the line of the collar or border piece with a No. 2 tube; overpipe it with a No. 1 and finish with a No. 1 tube placed alongside the other line. Finer linework is made by using No. 1 and No. 0 tubes instead of No. 2 and No. 1.

Designs for borders that can be traced from the book are given in Fig. 68.

Plaques

A plaque is the most useful form of decoration. It is simple to carry out; takes very little time and has many uses.

It is the plaque that comes to the rescue on occasions when someone requests an anniversary or birthday cake at very short notice. Perhaps you have some spare runout numerals in a box and a butterfly left over from a wedding cake, but would they look right on a 20 cm (8 in.) cake? Probably not, *on their own*, but quite possibly they could be used to advantage when placed on a plaque. Alone they would be too small and appear unrelated. Assembled on a plaque of a *suitable size for the size of the cake*, they could form a design.

To give another example, the Father Christmas in Fig. 74 measures 5 cm (2 in.) × 6 cm ($2\frac{1}{2}$ in.) and is intended for a 10 cm (4 in.) cake. Assembled on a plaque measuring 8 cm (3 in.) square, it becomes suitable for a 15 cm (6 in.) cake.

It can be seen that a supply of plaques of all shapes, sizes and colours is a valuable asset. The best way of achieving this is to have available several small boards on which tracings have been placed; covered with waxed paper and stuck down with tape. Stacked on a shelf or in a cupboard they will lie flat and be ready for use. Left over icing from borders or figures (but not from coating as this may contain glycerine) irrespective of colour or volume, need not be wasted. The preparation has already been made and the work will take very little time to complete.

Points to Remember

1. Small bags fitted with No. 1 tubes are used for outlining.
2. Outlines are made using the *touch, lift* and *place* method.
3. Large bags are used for flooding plaques.
4. Icing should be left to stand for several hours after it has been diluted. It must be stirred before use and re-checked for consistency.

(text continued p. 112)

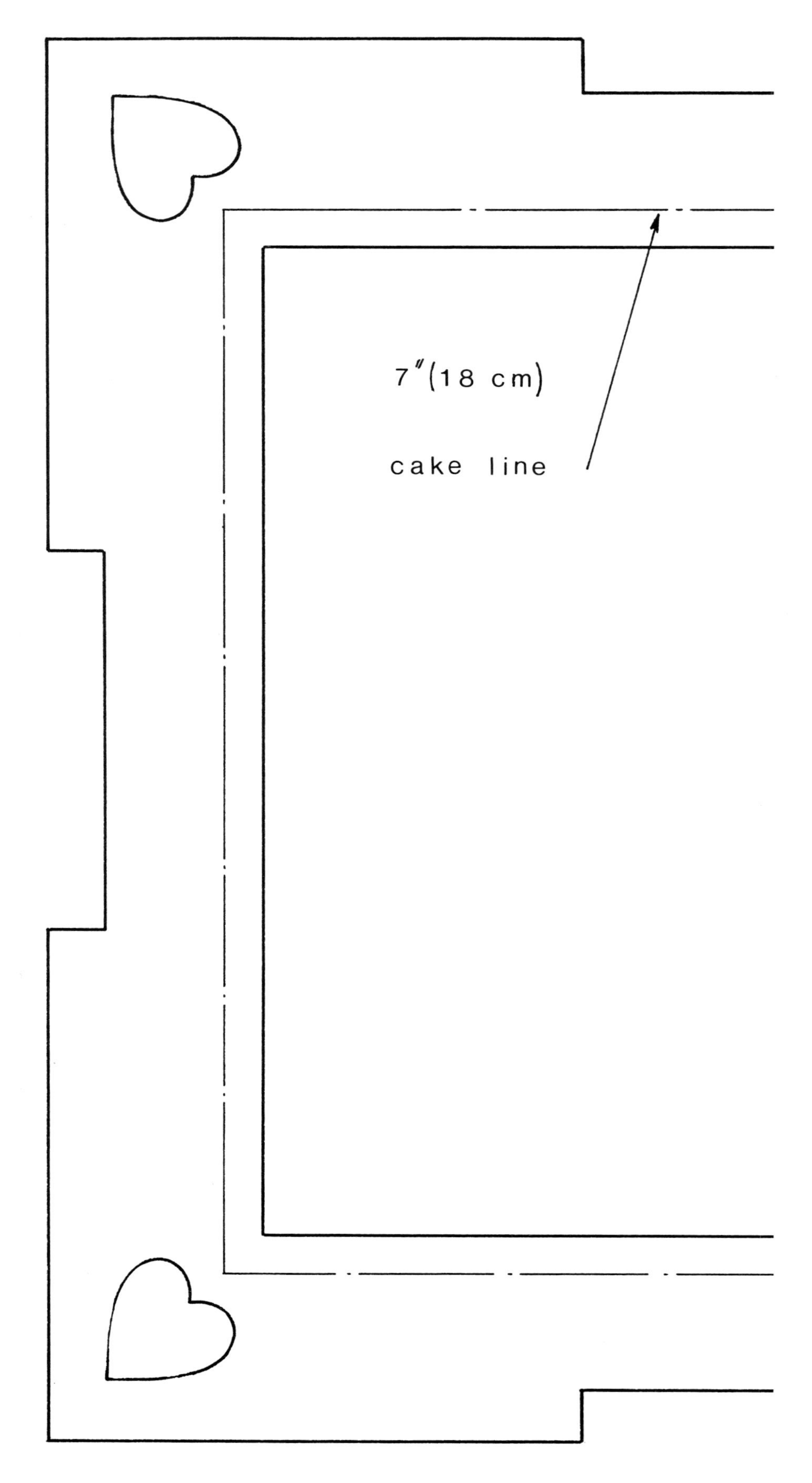

Fig. 68 Working drawings for runout borders. Half of pattern at full size for tracing. See opposite for whole border pattern.

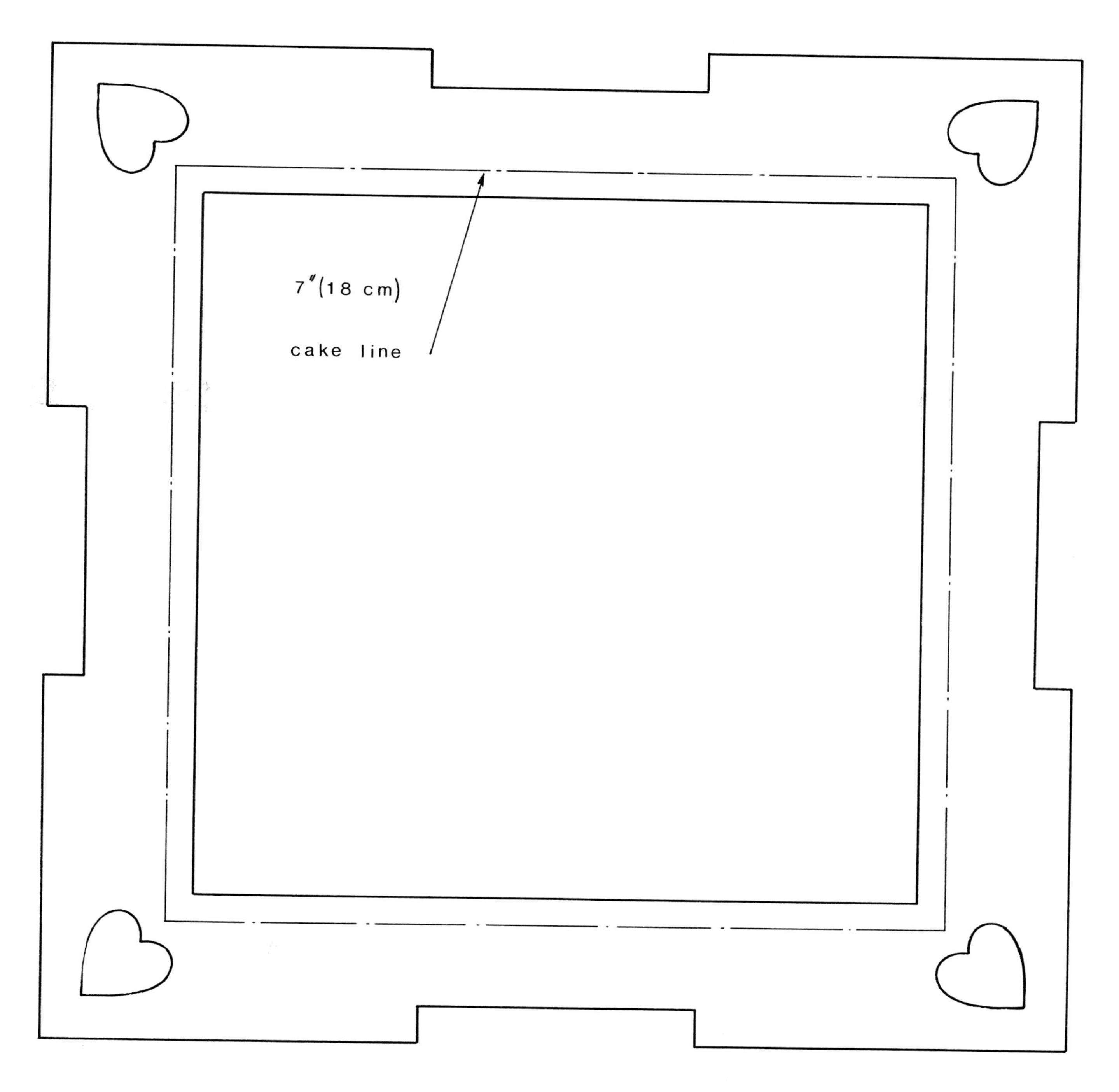

Fig. 68 — *contd.* Pattern reduced in size to show whole of border.

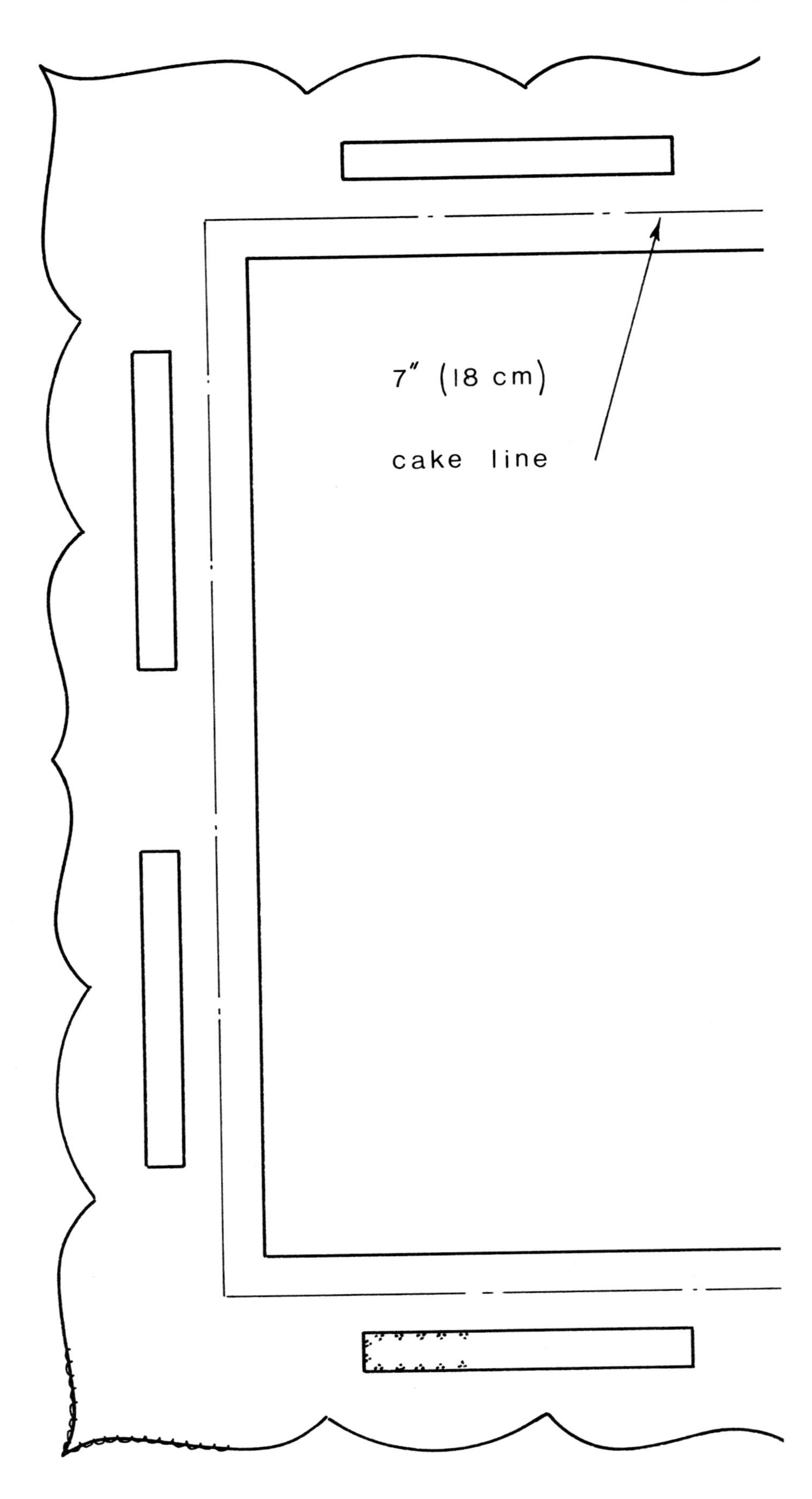

Fig. 68 — *contd.* Half of pattern at full size for tracing. See opposite for whole border pattern.

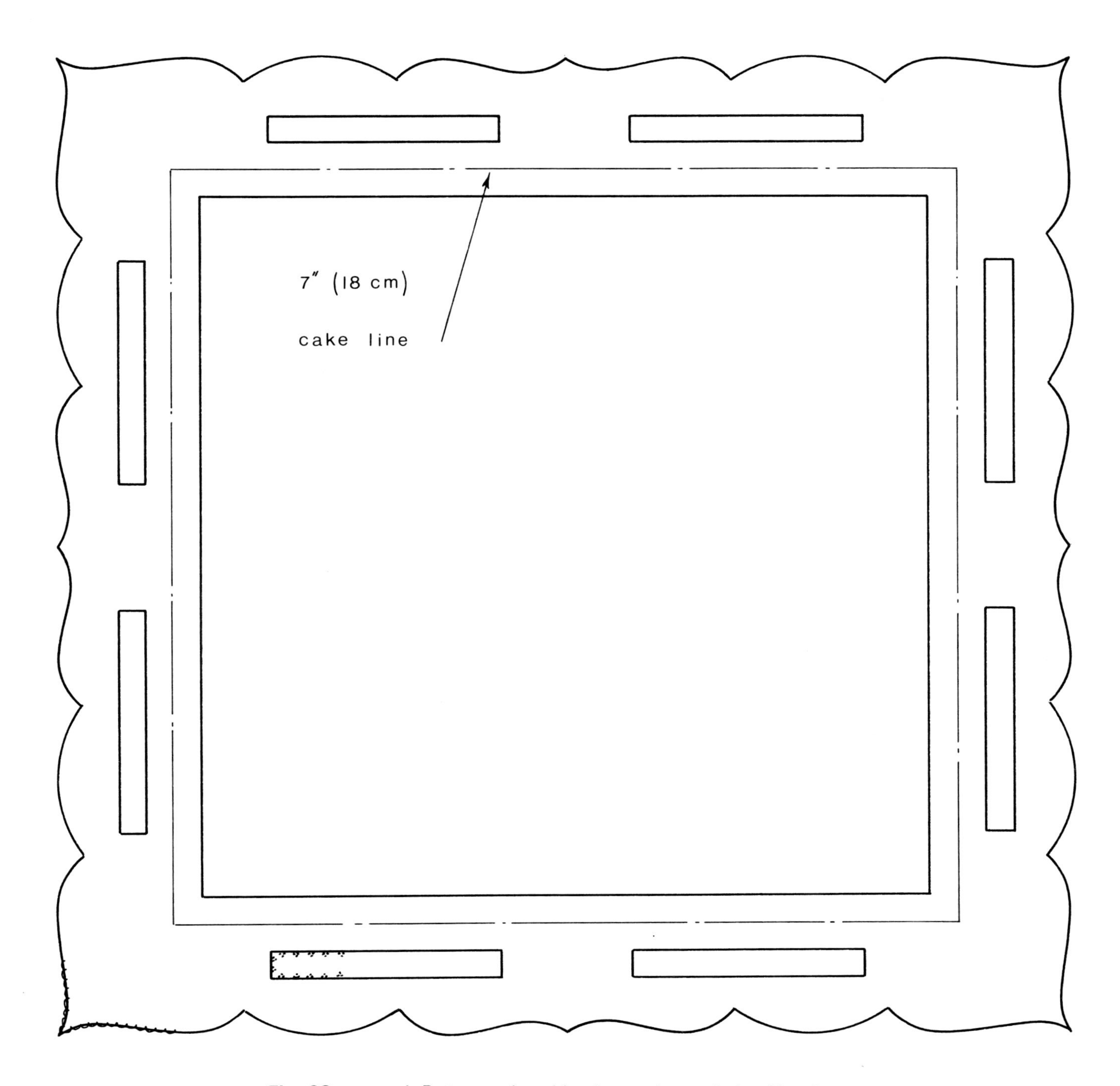

Fig. 68 — *contd.* Pattern reduced in size to show whole of border.

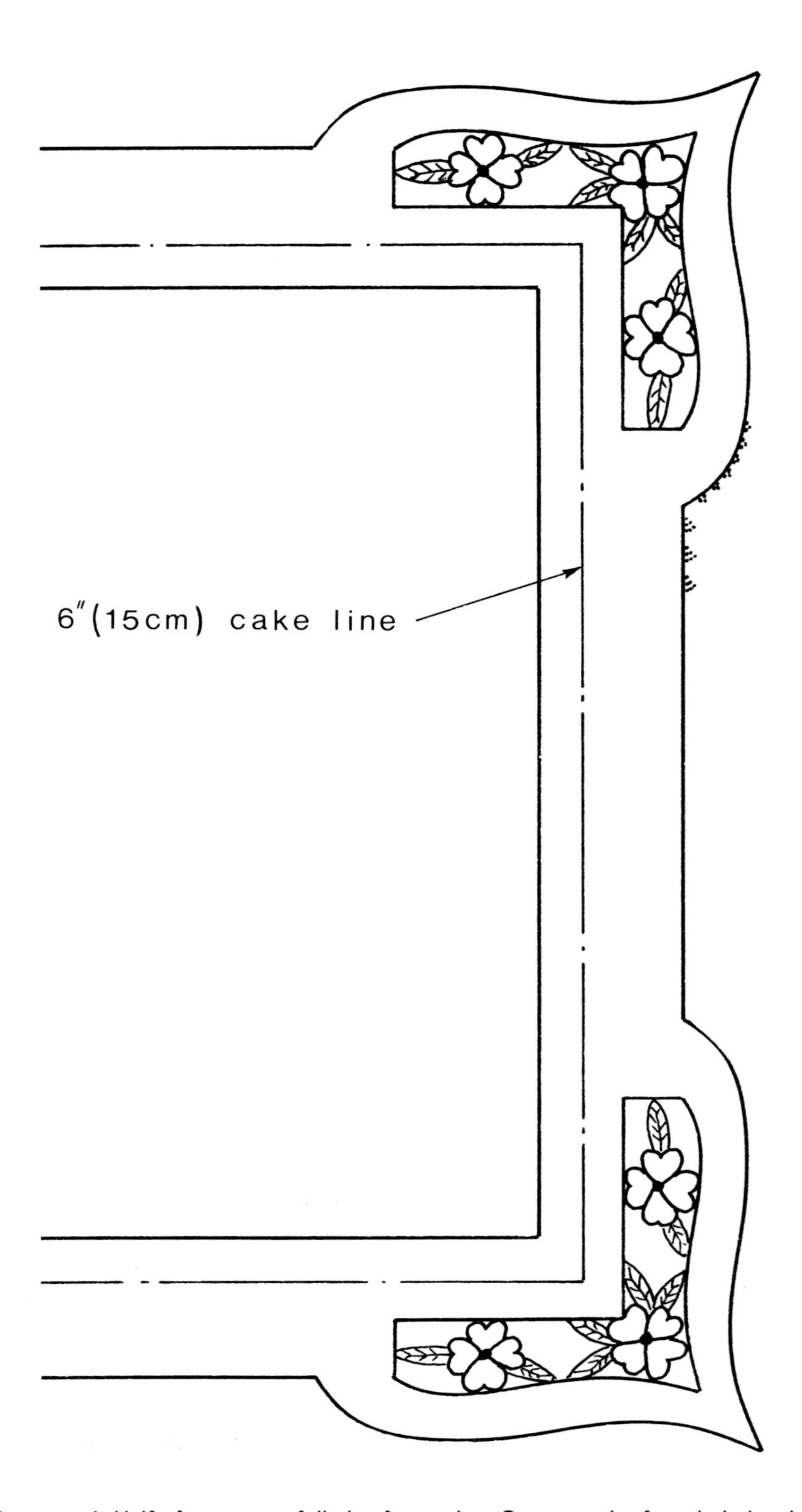

Fig. 68 — *contd.* Half of pattern at full size for tracing. See opposite for whole border pattern.

Fig. 68 — *contd.* Pattern reduced in size to show whole of border.

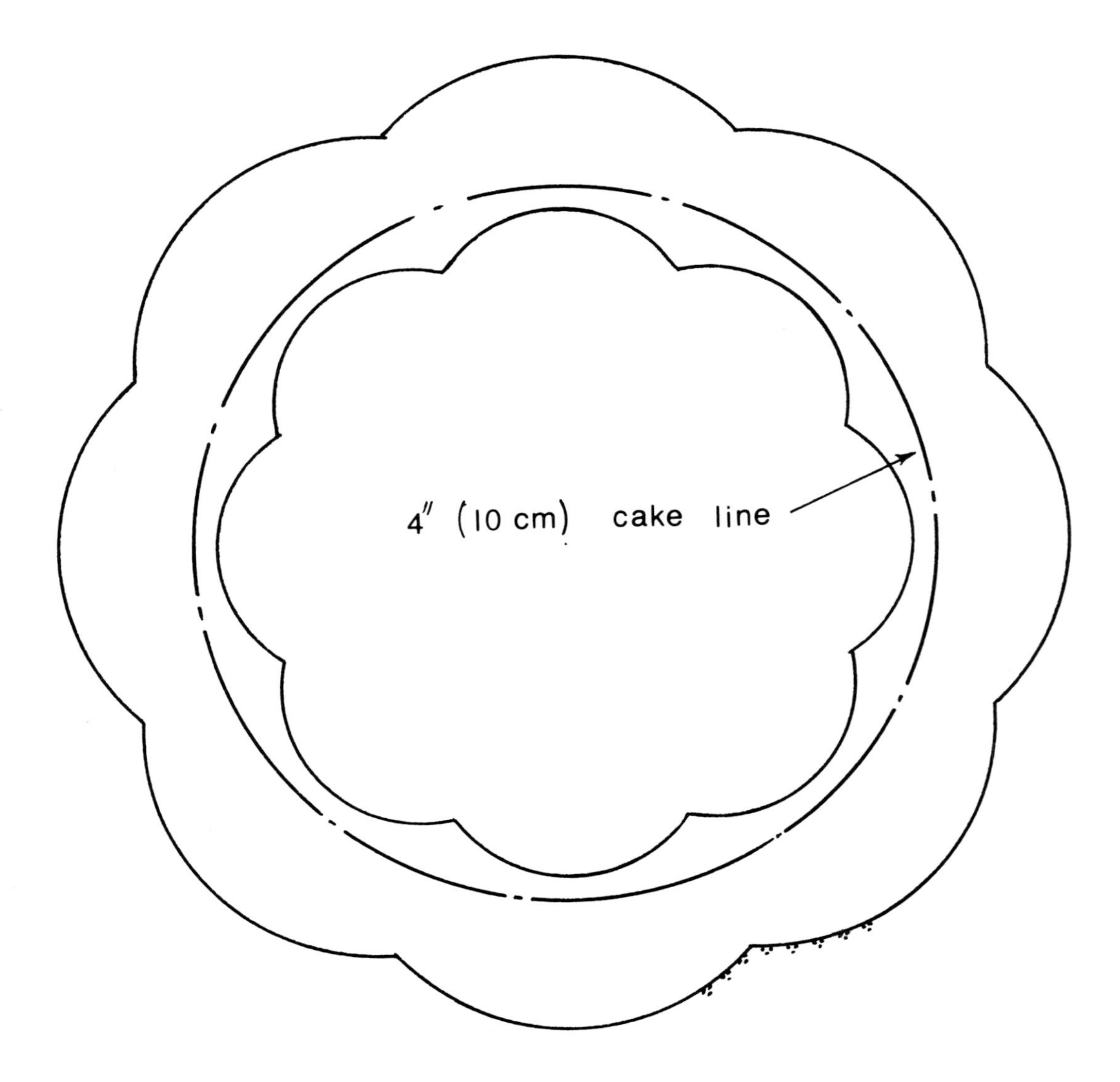

Fig. 68 — *contd.*

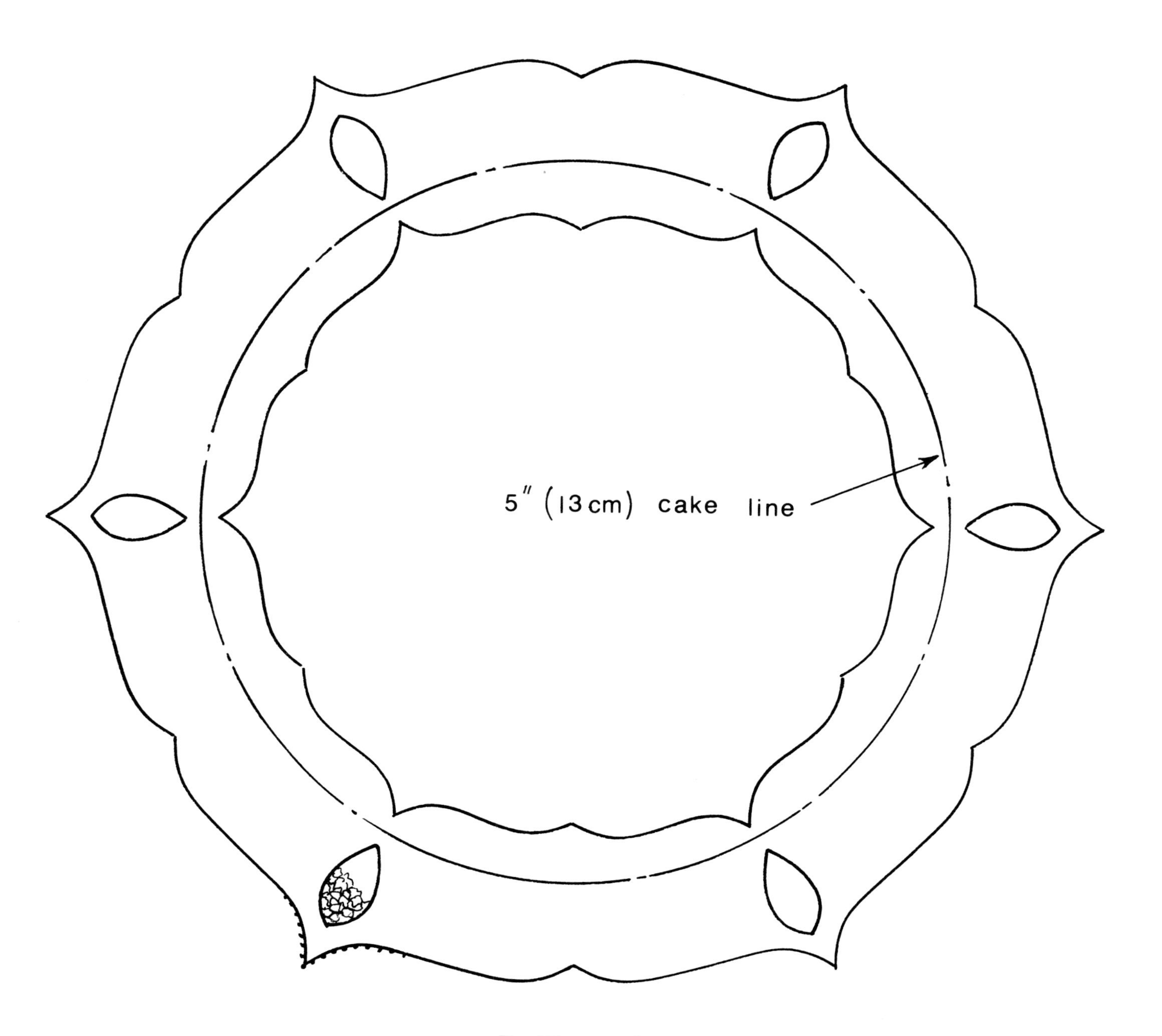

Fig. 68 — *contd.*

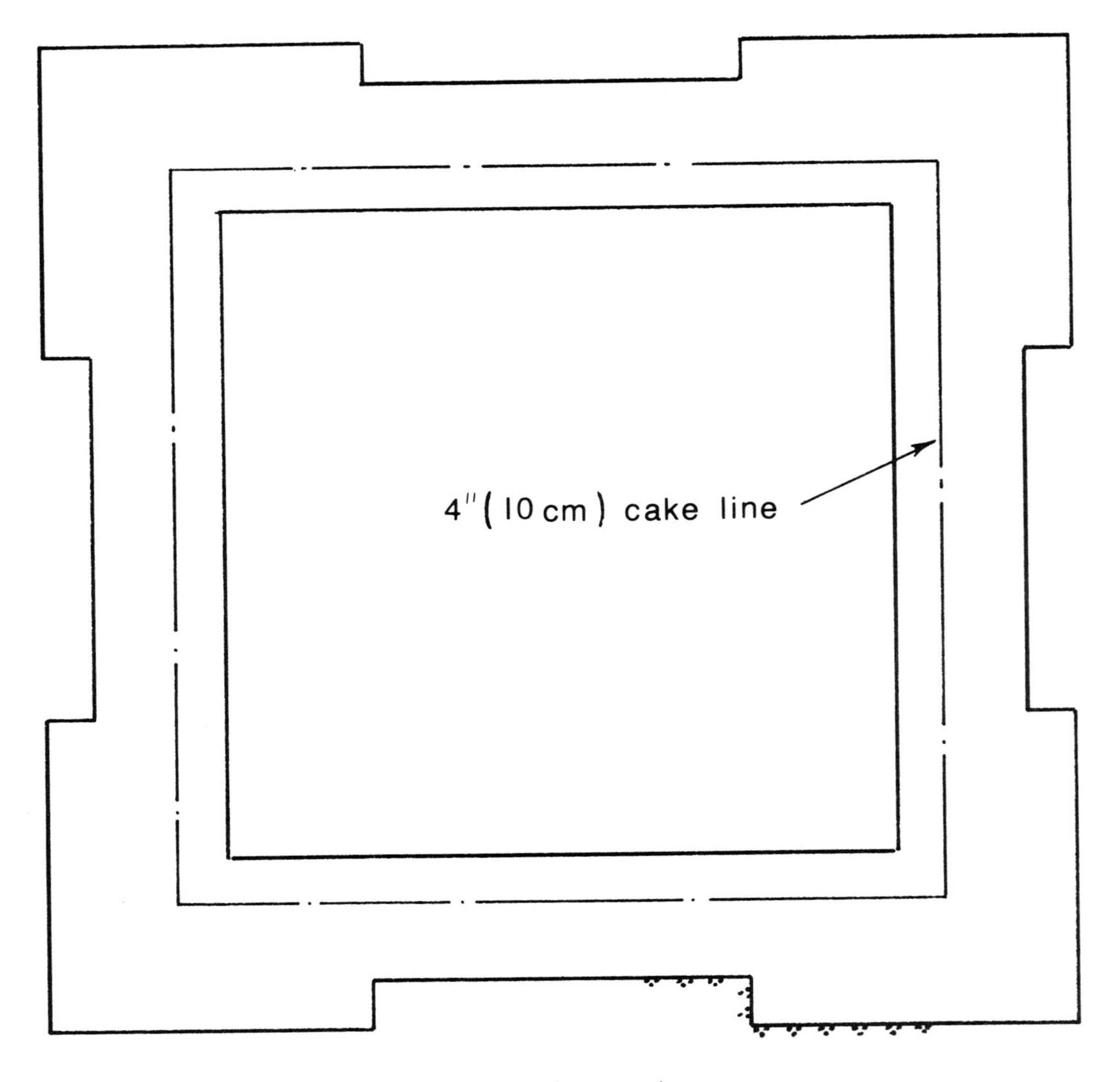

Fig. 68 — *contd.*

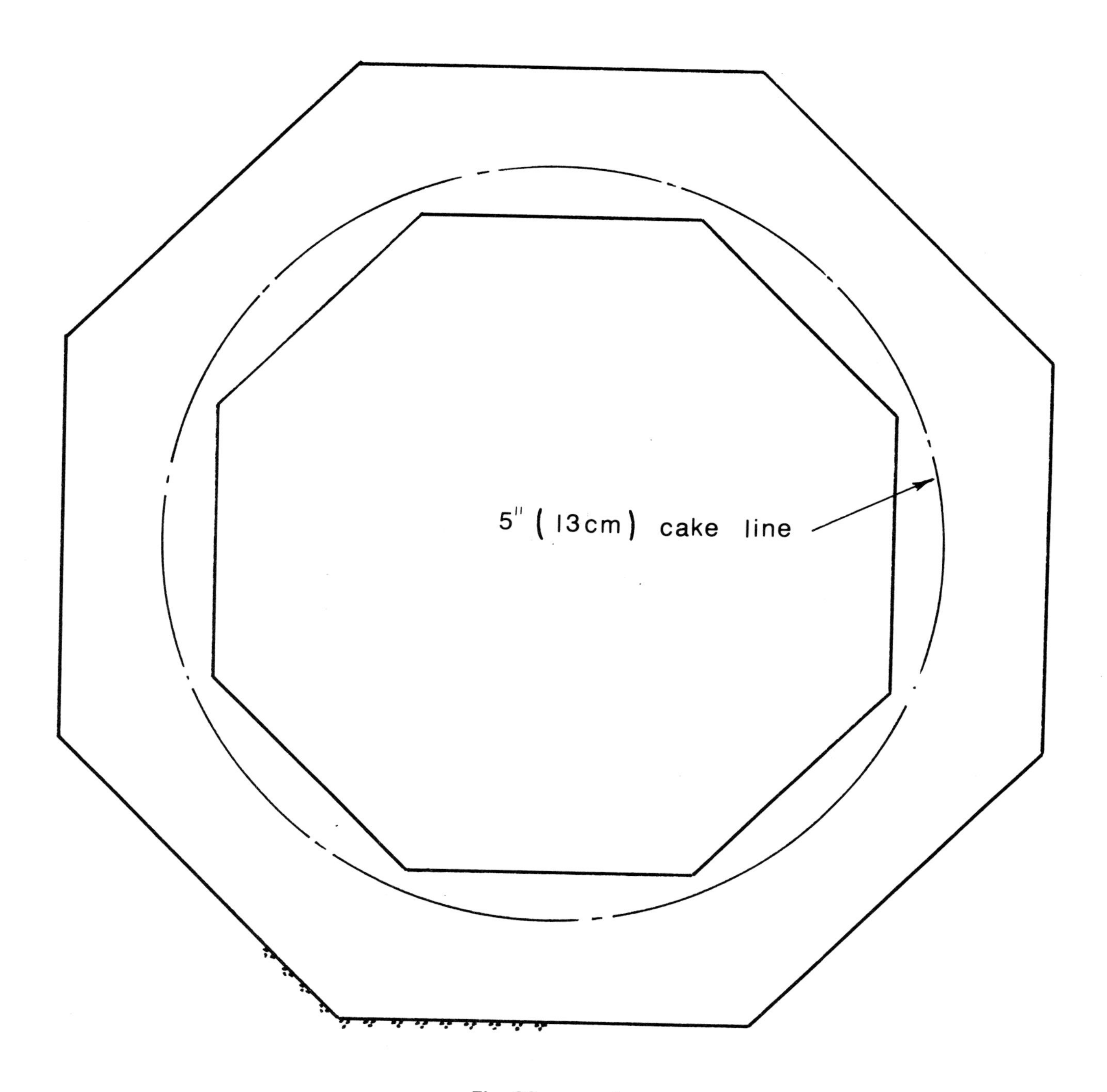

Fig. 68 — *contd.*

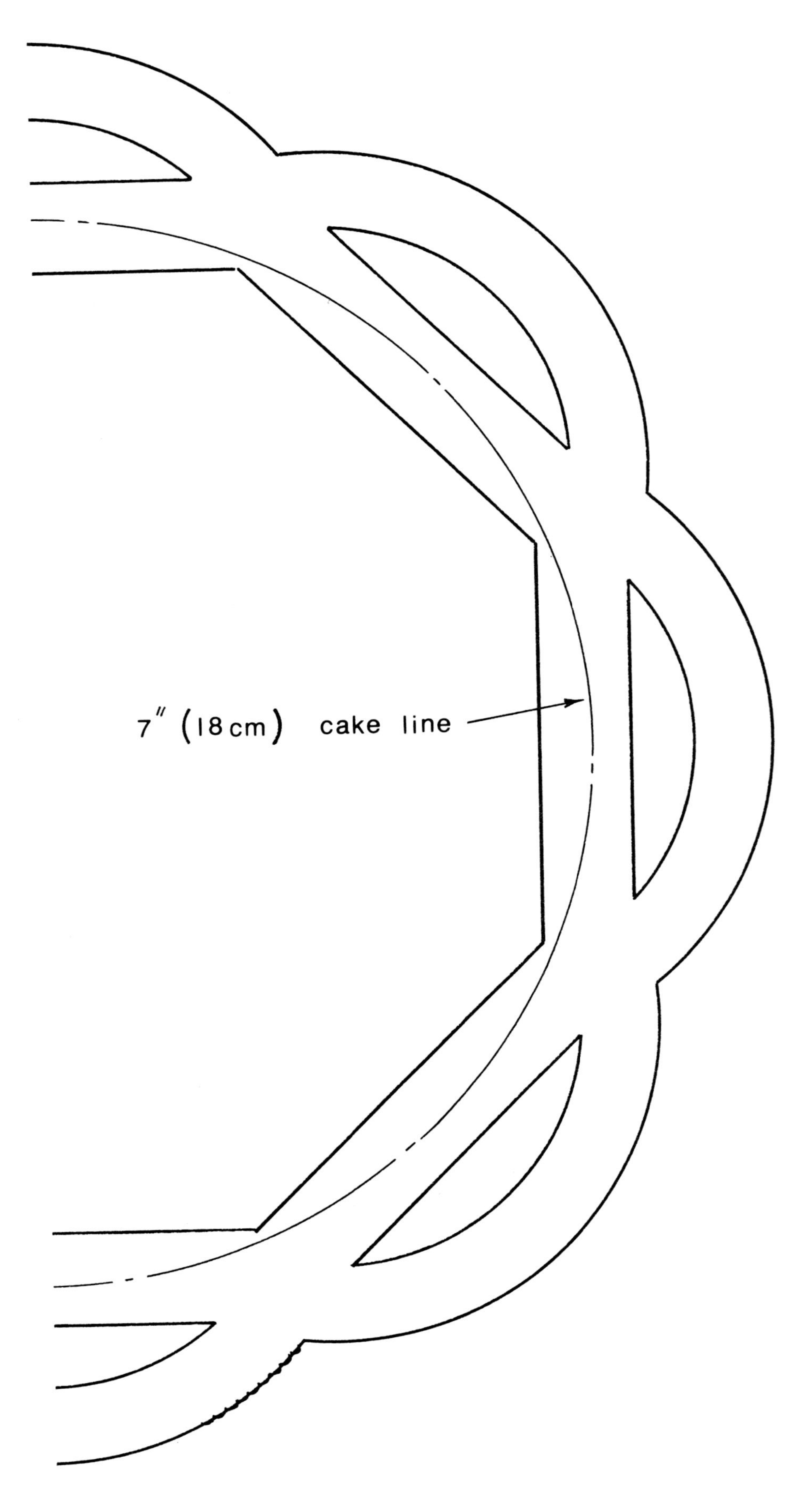

Fig. 68 — *contd.* Half of pattern at full size for tracing. See opposite for whole border pattern.

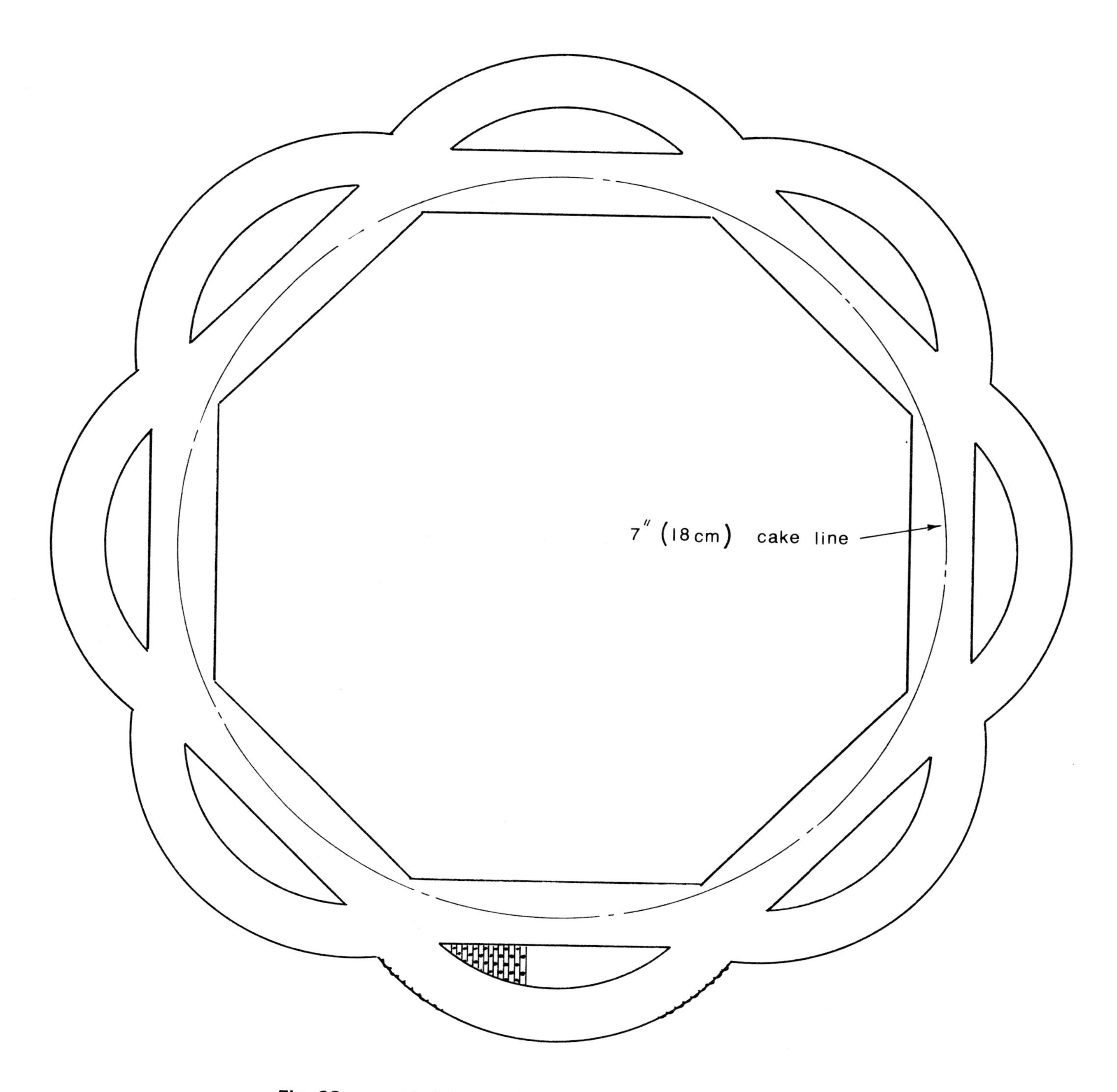

Fig. 68 — *contd.* Pattern reduced in size to show whole of border.

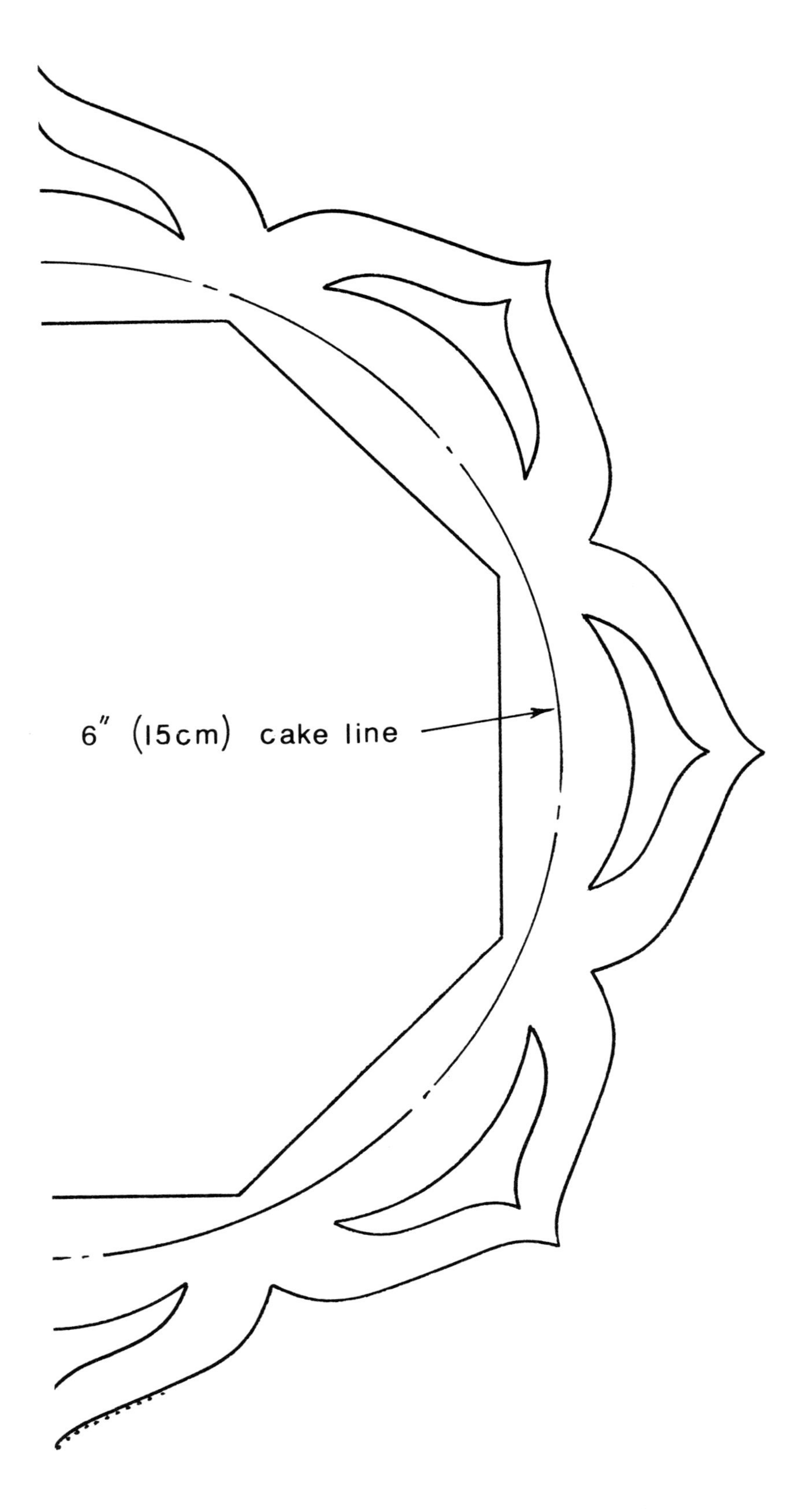

Fig. 68 — *contd.* Half of pattern at full size for tracing. See opposite for whole border pattern.

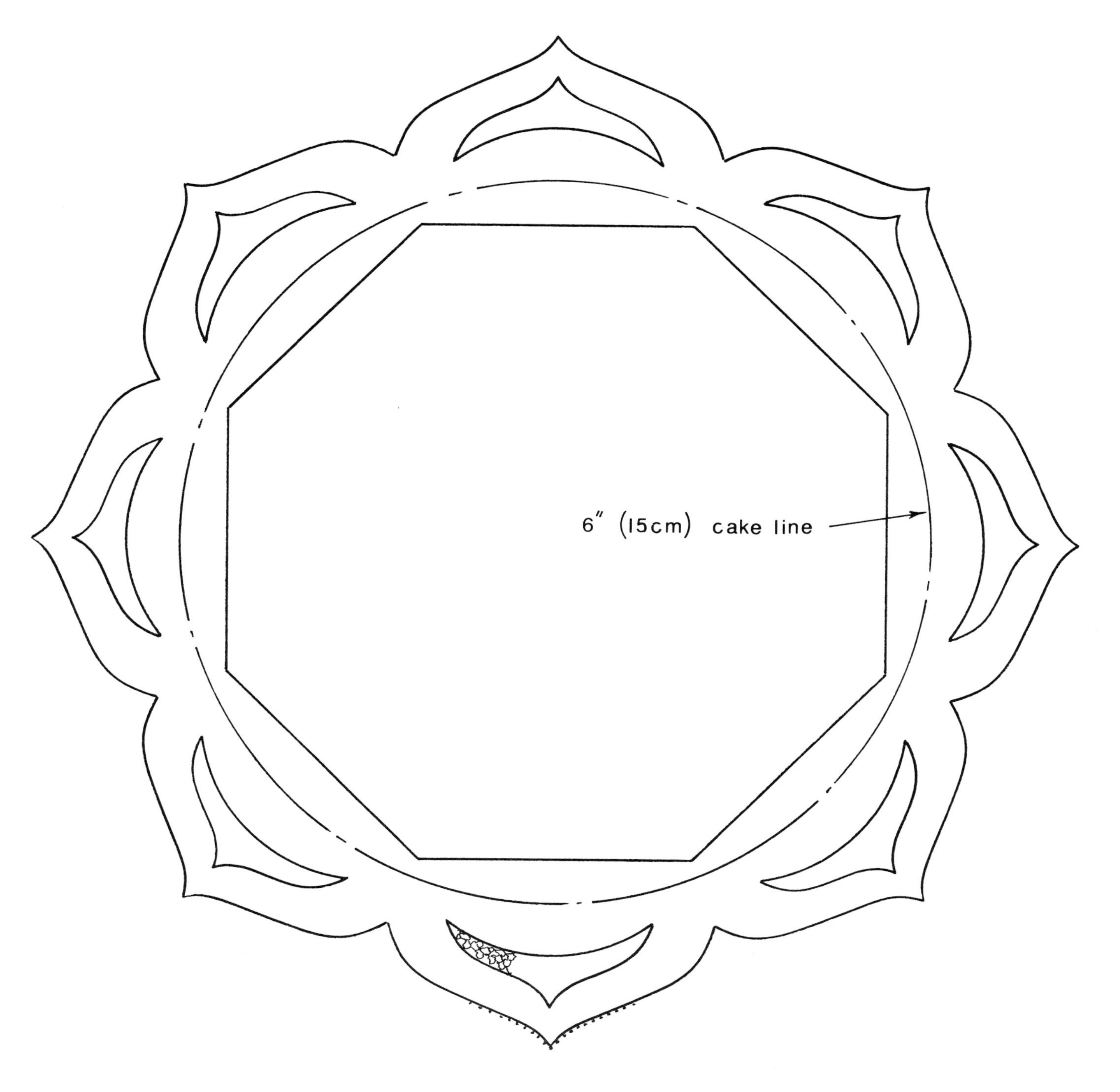

Fig. 68 — *contd.* Pattern reduced in size to show whole of border.

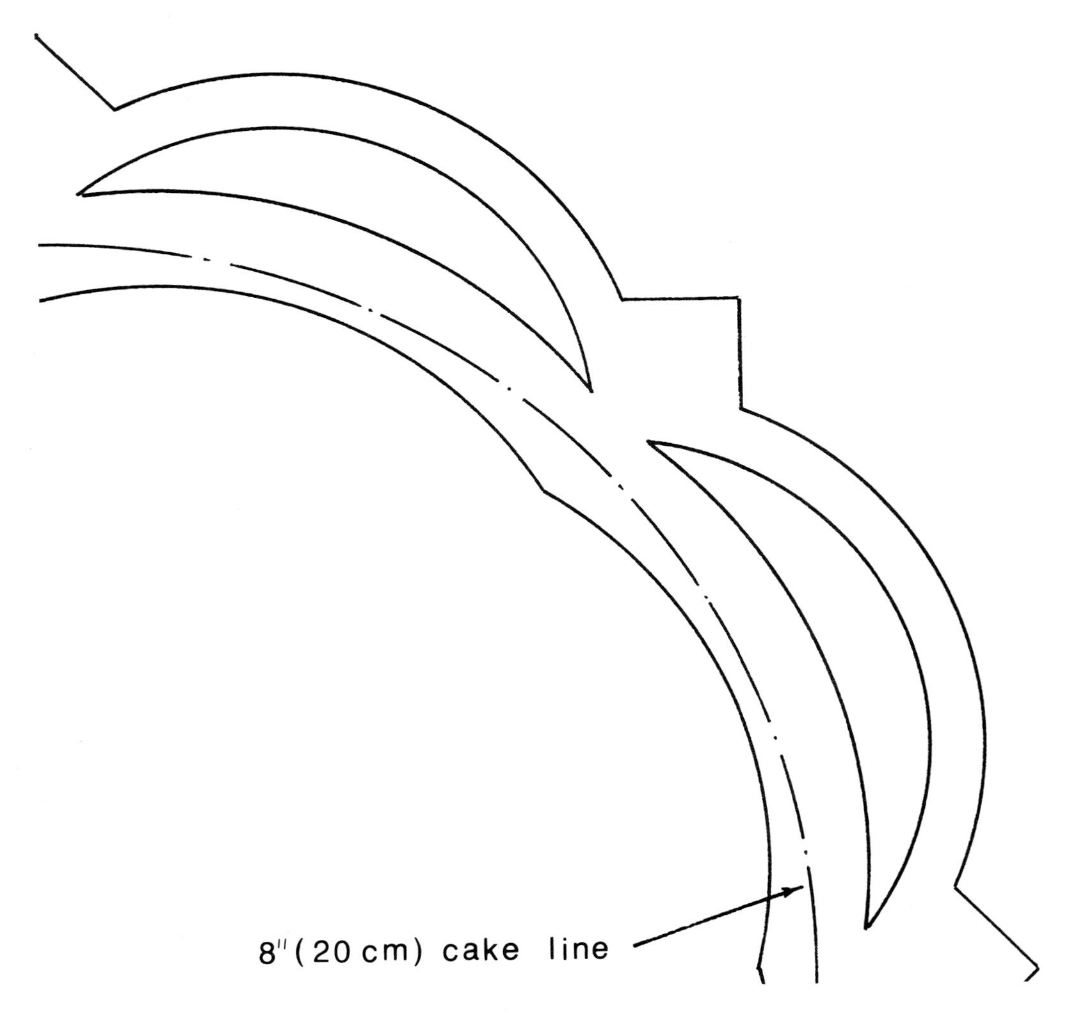

Fig. 68 — *contd.* Quarter of pattern at full size for tracing. See opposite for whole border pattern.

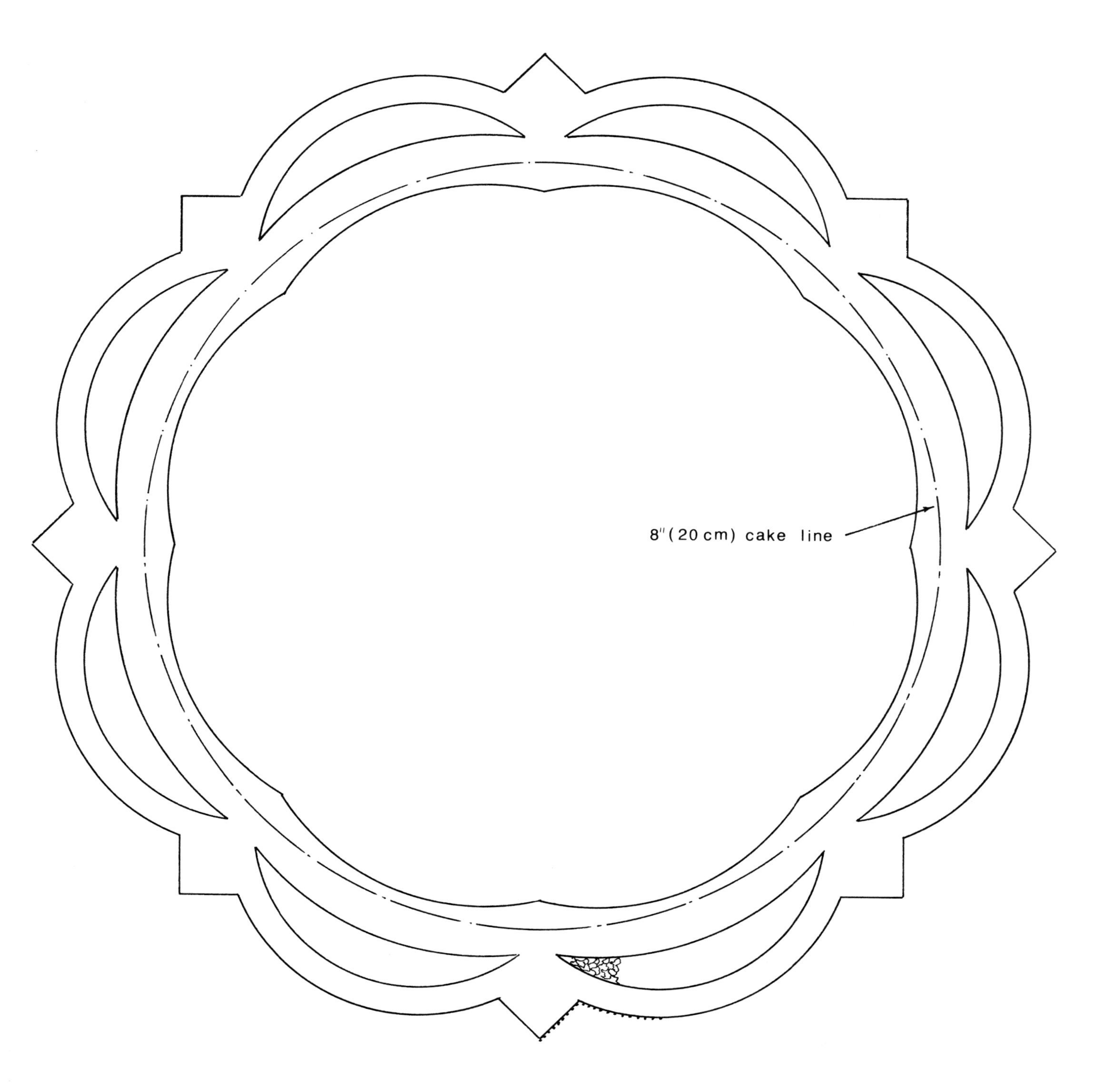

Fig. 68 — *contd.* Pattern reduced in size to show whole of border.

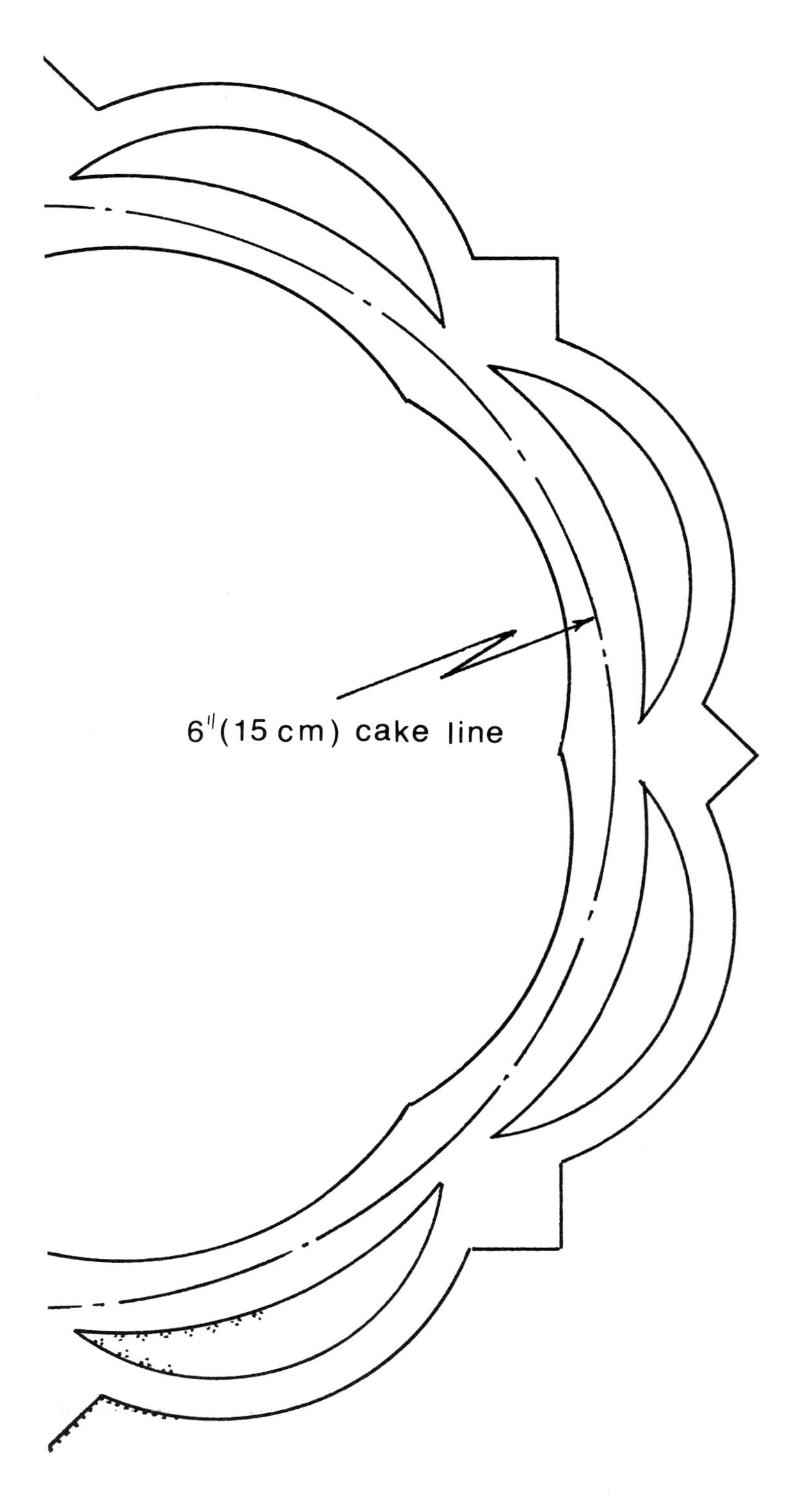

Fig. 68 — *contd.* Half of pattern at full size for tracing. See opposite for whole border pattern.

Fig. 68 — *contd.* Pattern reduced in size to show whole of border.

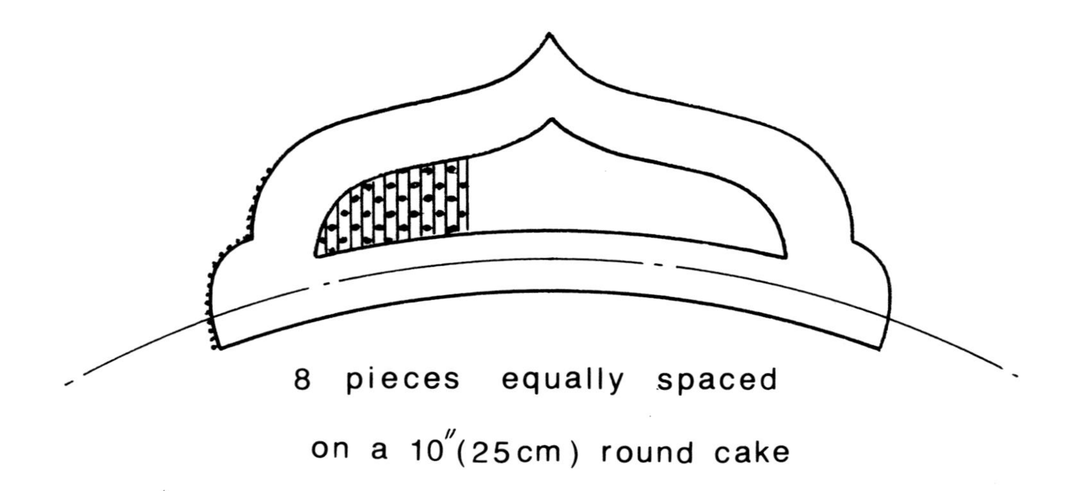
8 pieces equally spaced
on a 10"(25cm) round cake

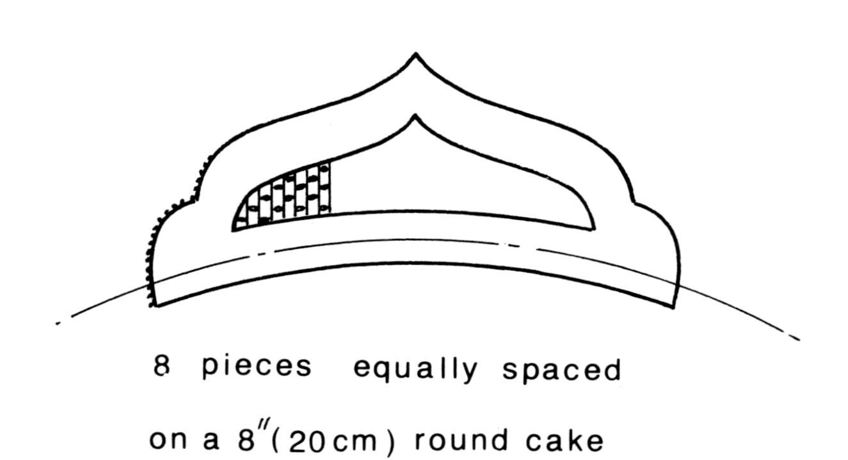
8 pieces equally spaced
on a 8"(20cm) round cake

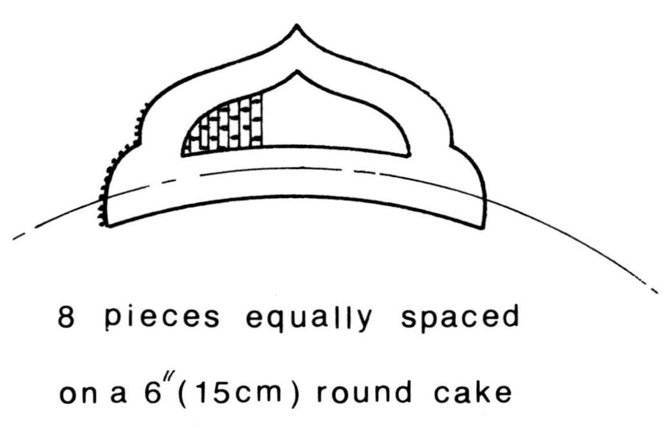
8 pieces equally spaced
on a 6"(15cm) round cake

Fig. 68 — *contd.*

10 pieces equally spaced

on a 11″(28cm) round cake

8 pieces equally spaced

on a 8″(20 cm) round cake

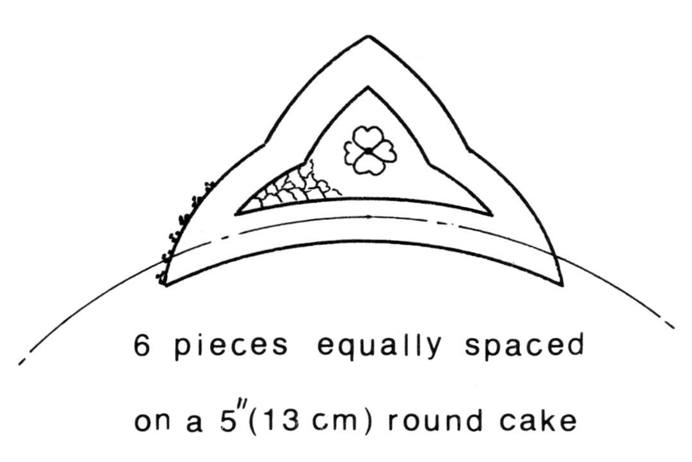

6 pieces equally spaced

on a 5″(13 cm) round cake

Fig. 68 — *contd.*

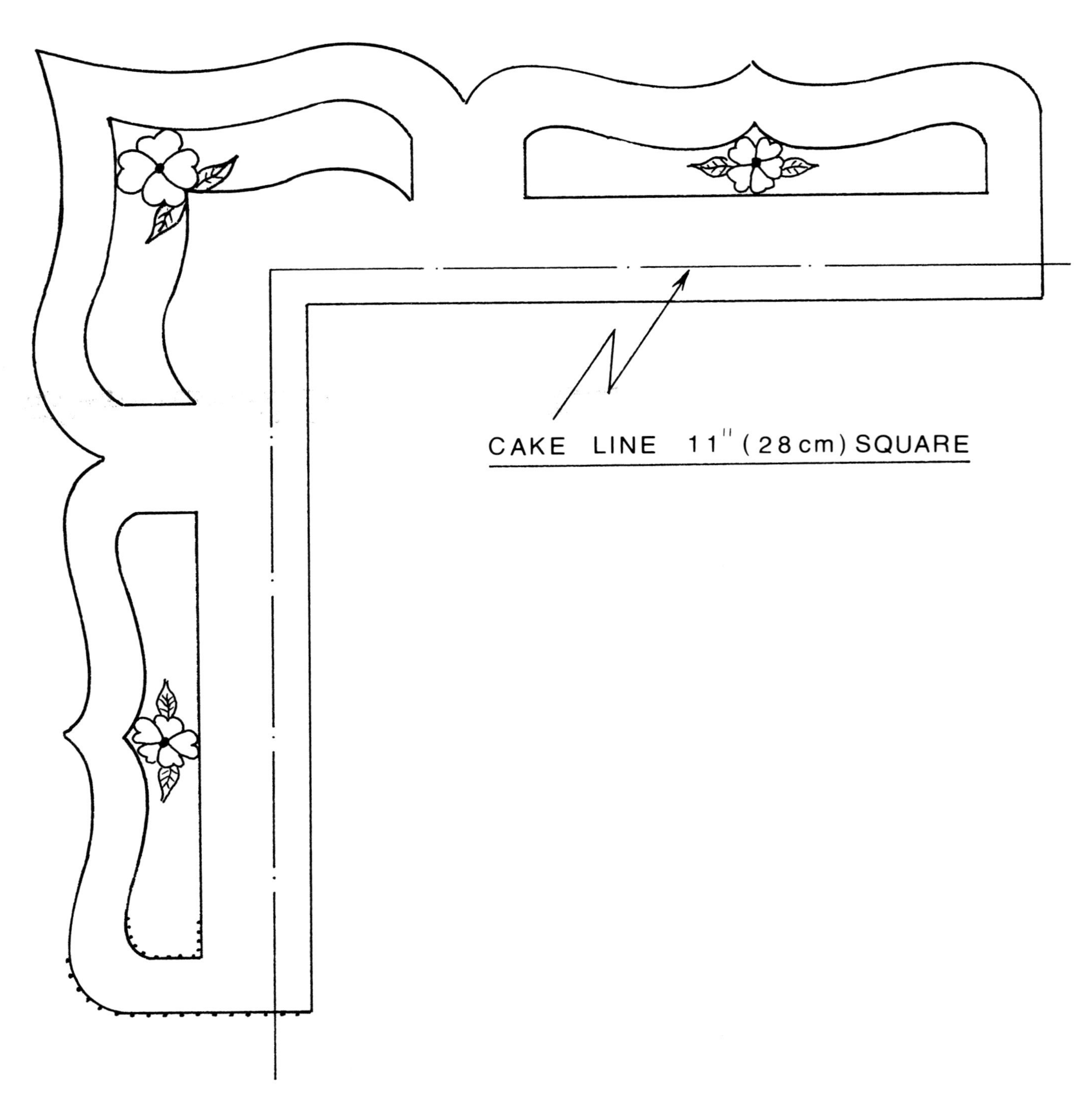

Fig. 68 — *contd.*

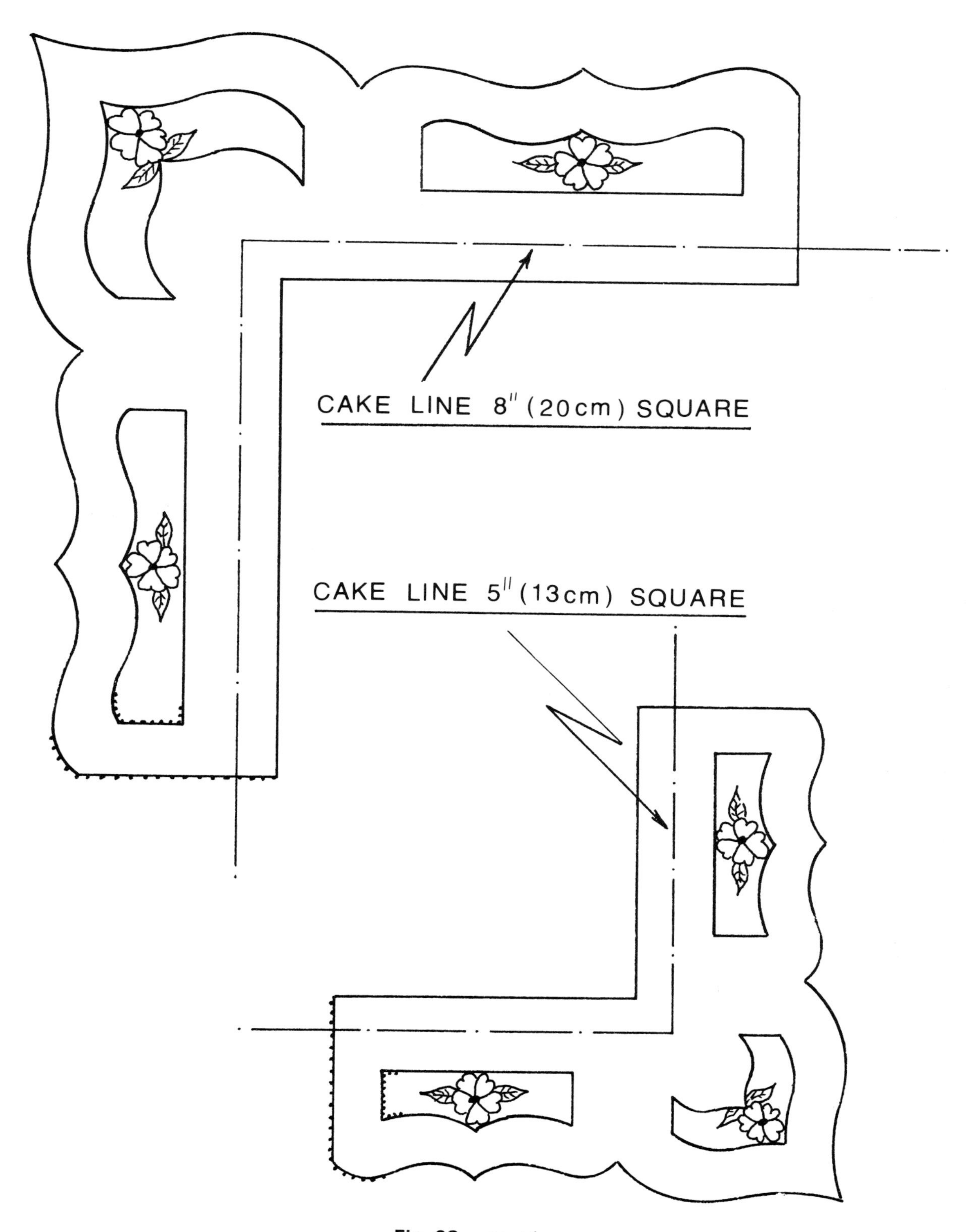

Fig. 68 — *contd.*

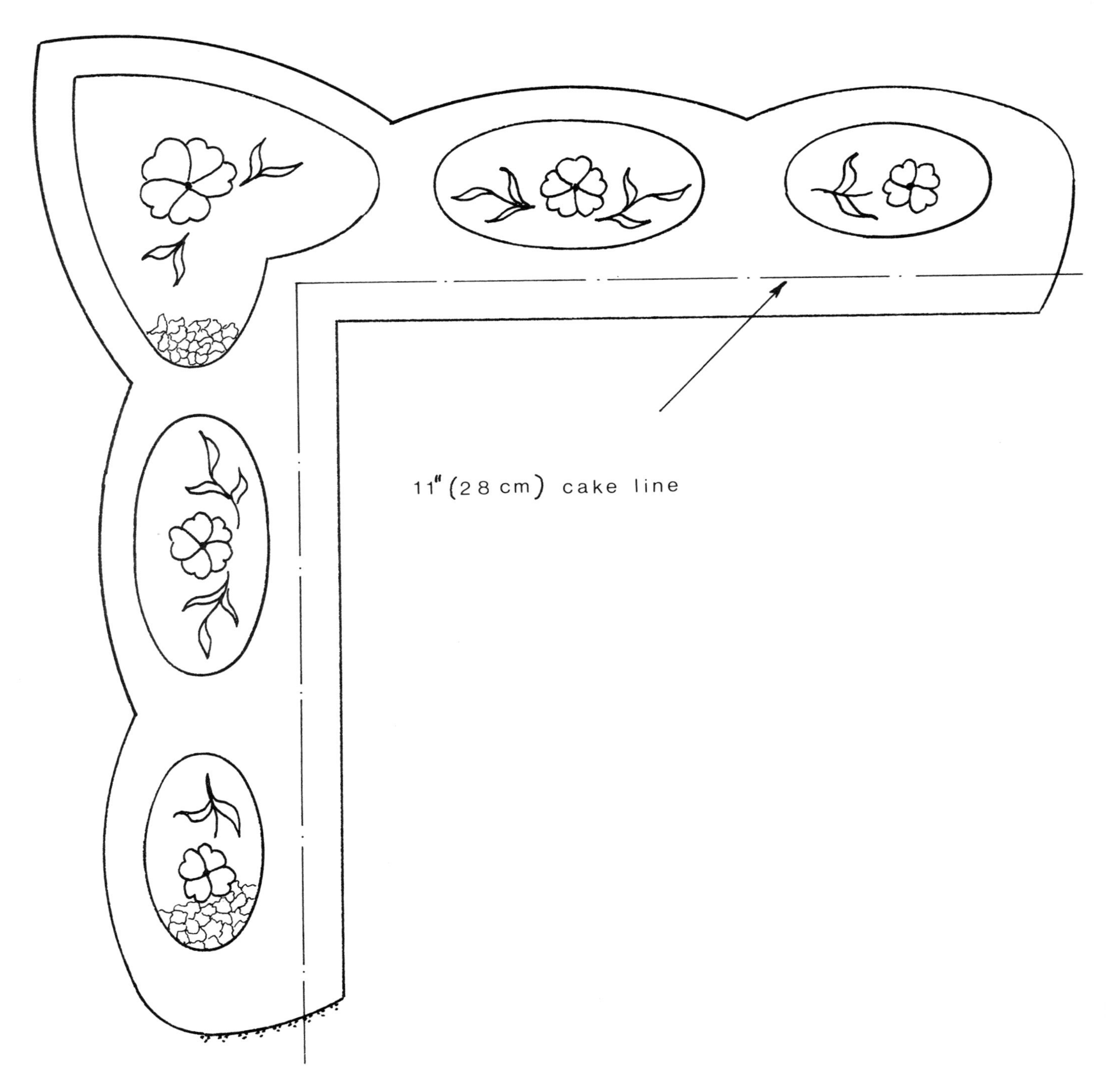

Fig. 68 — *contd.*

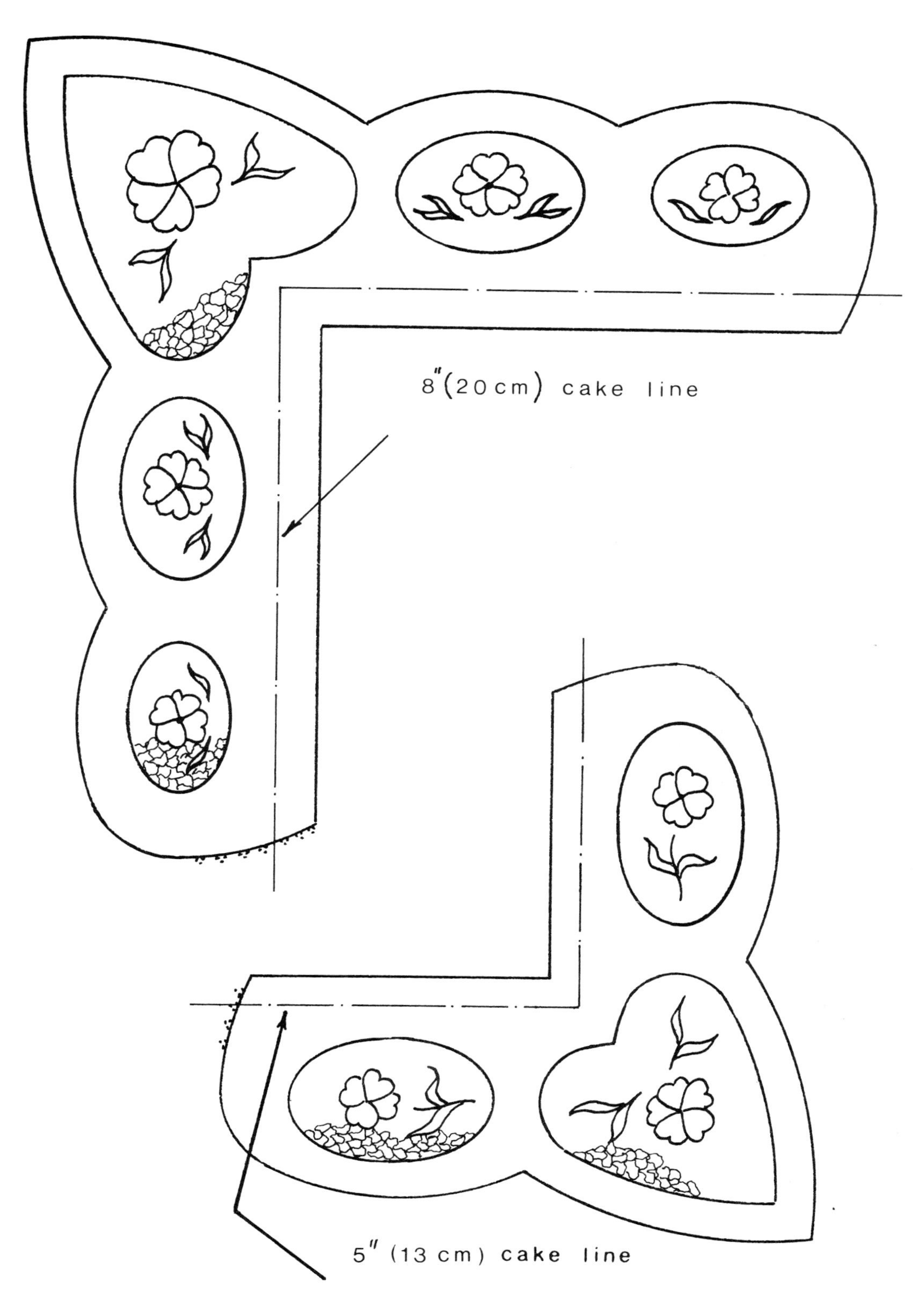

Fig. 68 — *contd.*

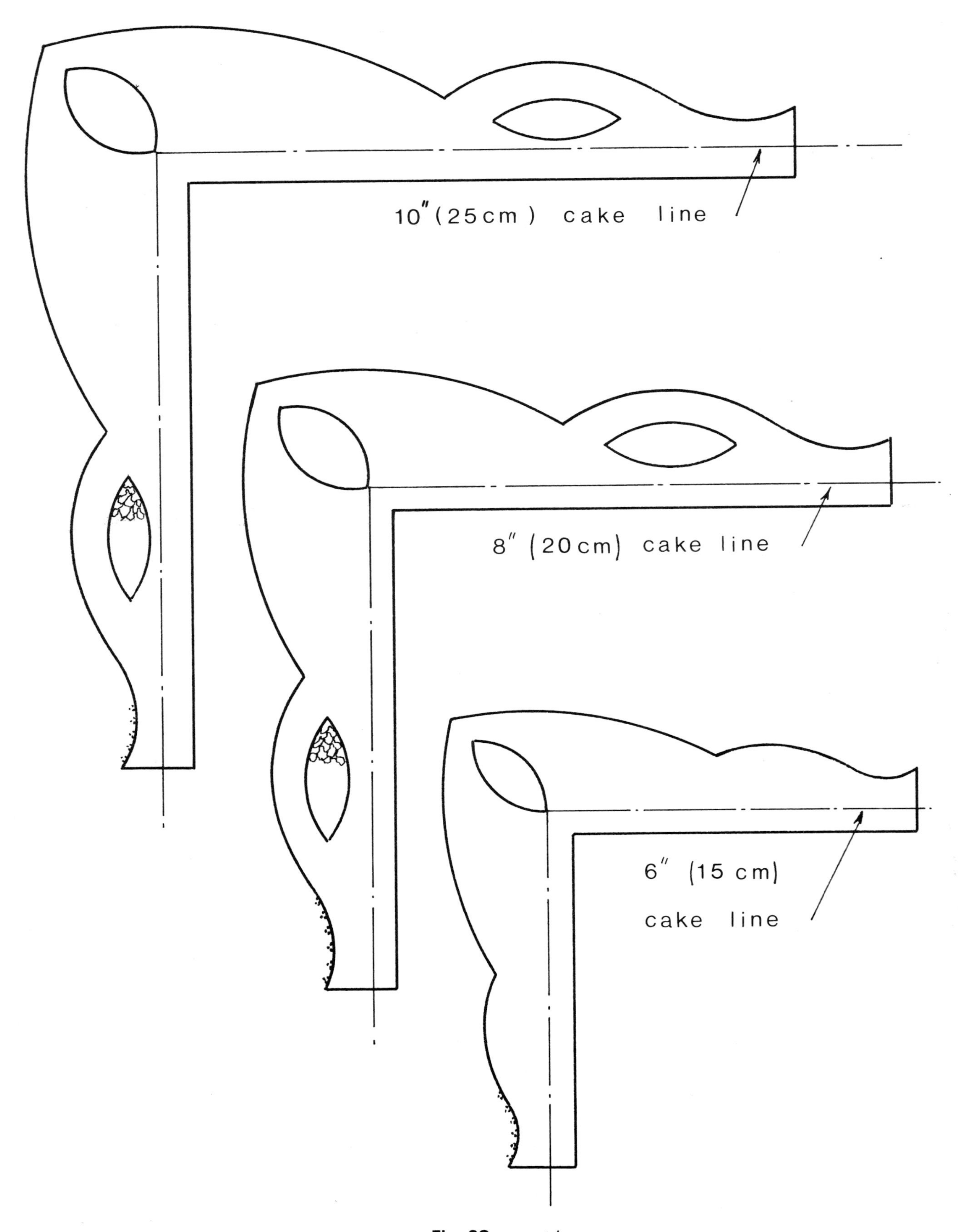

Fig. 68 — *contd.*

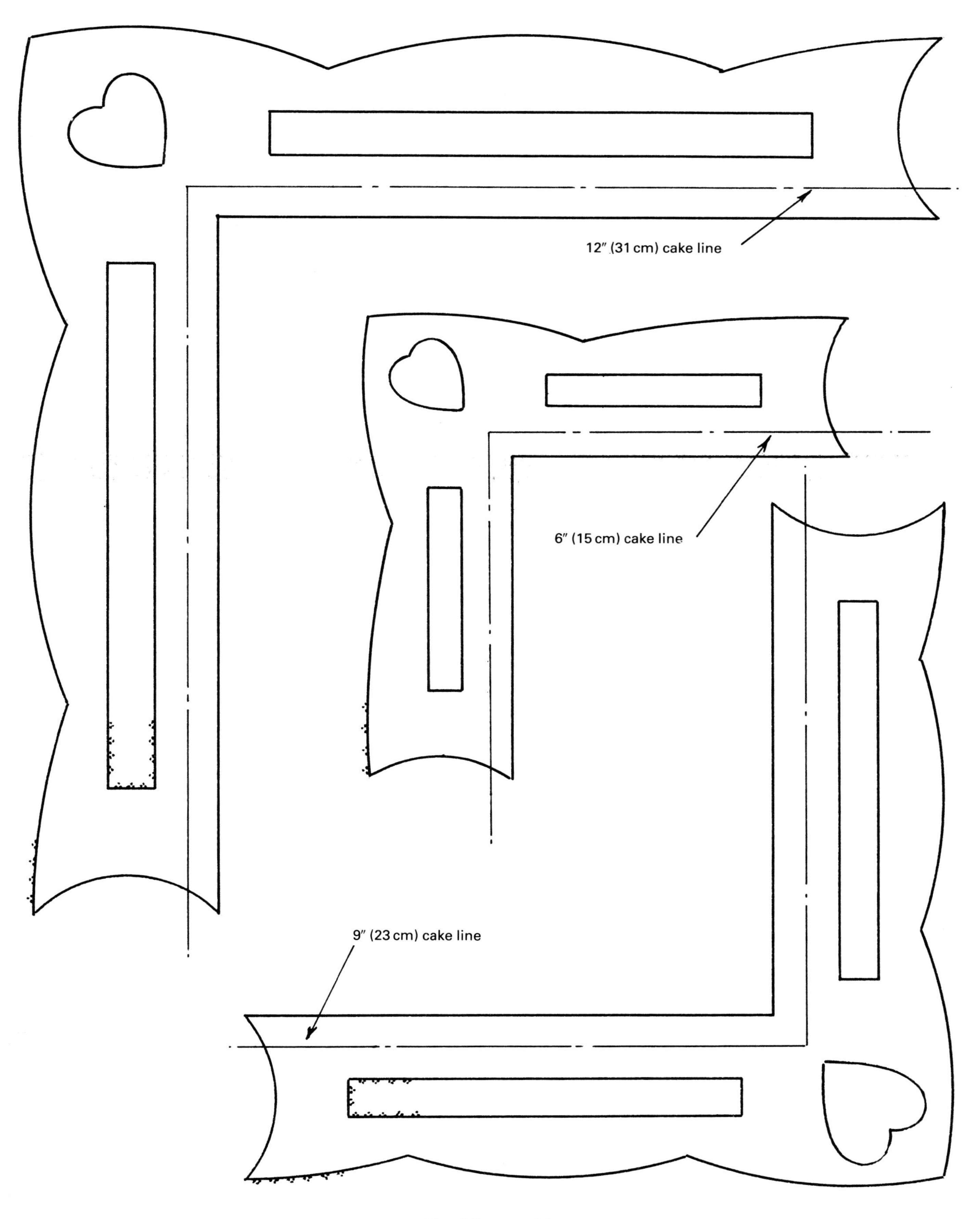

Fig. 68 — *contd.*

5. Plaques should, whenever possible, be placed on individual boards.

6. In order that the outline is not visible on the finished plaque, it should be flooded before the outline has had a chance to crust.

7. If a painted edge or an edge of contrasting colour to that of the plaque is desired, the outline *is* allowed to crust before painting and flooding.

8. Unless a contrasting outline is required, the icing must be exactly the same colour as that of the flooding.

9. The hole in the bag will depend upon the size of the plaque to be flooded; ie. the smaller the plaque, the smaller the hole and never larger than a No. 3 tube.

10. A smooth, rounded plaque is achieved by flooding thickly and quickly with icing of the correct consistency. It should be soft but not runny and brushed *on but not over* the outline.

11. Do not apply icing directly to the outline as the weight of the icing may cause it to break.

12. To obtain a shiny finish, leave under a lamp for as long as possible before transferring to a warm place to dry.

Procedure for Runout Plaques

1. Have ready large bags of softened icing of the same consistency as that used for runout borders.

2. Pipe an outline; cut a hole in the bag and flood a line around the inner edge of the plaque almost, but not quite to the linework. Use a paint brush to take the icing onto but not over the outline, as quickly as possible. Flood the plaque thickly but evenly, moving the bag from side to side and making the icing smooth with the aid of a paint brush. The paint brush may also be used to prick any air bubbles which may have appeared (Fig. 69).

3. If the icing is not absolutely smooth, lift the board and *gently* tap the underneath with the fingers. This must be done before the icing has formed a skin. It must not be attempted if there is more than one plaque on the board.

4. Place under a lamp whilst outlining and flooding the next plaque. Keep plaques under a lamp for as long as possible to ensure a pleasing gloss. Leave in a warm atmosphere for several days and when dry store, still attached to the waxed paper, until required.

Decorating Plaques

Plaques can be decorated in numerous ways: painted, using edible colours; brushed with petal dust using a dry paint brush or cotton wool; piped with fine tubes or used to hold runout figures, monograms etc. It is possible to combine all these methods on a single plaque!

It is possible to trace a design on to a plaque prior to piping but this must be done very carefully in order not to break the plaque. Usually only part of the design is required to be traced to act as a guide for the remainder of the decoration. This point is illustrated in Figs 70–72.

1. The basic shape of the branches was traced from the greetings card and transferred to the plaque with a pencil (Fig. 70).

Fig. 69. Partially flooded plaque.

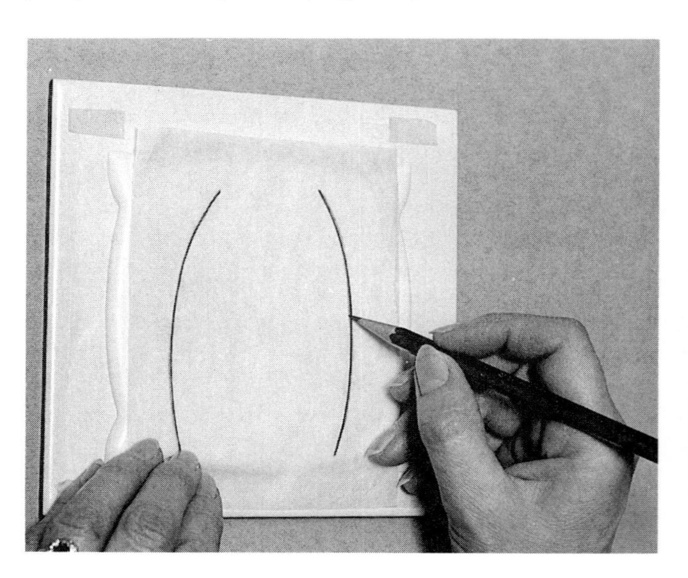

Fig. 70. Tracing on plaque.

2. When the paper was removed, a faint pencil line was visible and the line was overpiped with brown icing in a No. 0 tube using the *touch, lift* and *place* method (Fig. 71).

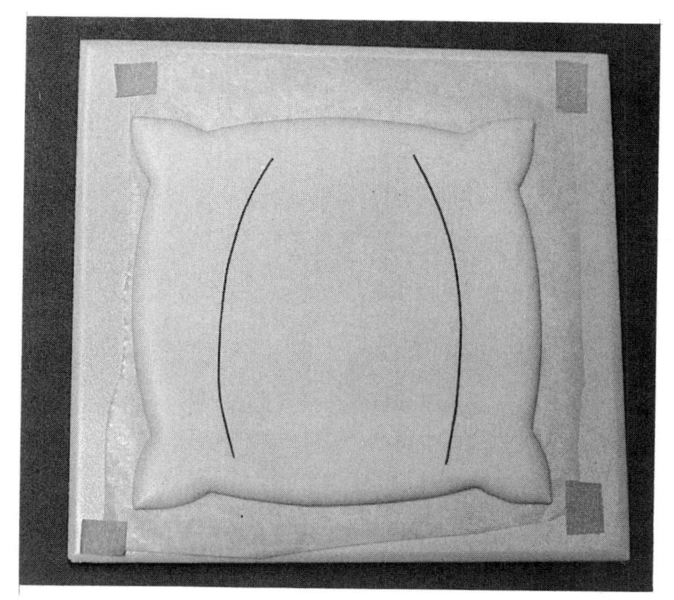

Fig. 71. Lines piped on plaque.

3. Additional branches and green leaves were added with No. 0 tubes but with the tubes touching the surface of the plaque whilst piping (Fig. 72).

Fig. 72. Additional piping on plaque.

4. Flowers and additional foliage were piped after the runout figure was placed on to the plaque (Fig. 73).

This method can also be used when placing a runout directly on to the cake (Fig. 50).

Some form of edging is required and this is piped with a No. 0 tube before the plaque is placed on the cake. An exception to this is where small bulbs are used. These are piped *after* the plaque has been assembled as the bulbs are attached to the plaque and to the cake.

Placing Plaques on Cakes

1. Plaques are not placed on to cakes until the coating is dry.
2. Edging of plaques is carried out with a No. 0 tube.
3. Plaques intended for the top of cakes usually occupy a central position as do those for the sides of square cakes. Plaques for the sides of round cakes are positioned after the top border has been assembled and evenly spaced in line with the design on the top border.
4. The waxed paper is peeled from the plaque and a little icing placed on the back before it is stuck on to the cake. The icing will take a few moments to dry and whilst there is still time to make adjustments, a ruler should be used to ascertain that the plaque is in the correct position.
5. If an edging of small bulbs is desired, it is made after the plaque has been secured. The edging should touch the plaque and also make contact with the surface of the cake.
6. It is possible to secure a plaque with the aid of small bulbs only. The advantage of this method is that the plaque can be removed, intact when the bulbs are released. The disadvantage is that great care is required to ensure the plaque does not move whilst the edging is piped.

Photographs of decorated plaques are to be found in Figs 73 and 74.

Figs 47 and 75 contain drawings for plaques that can be traced from the book.

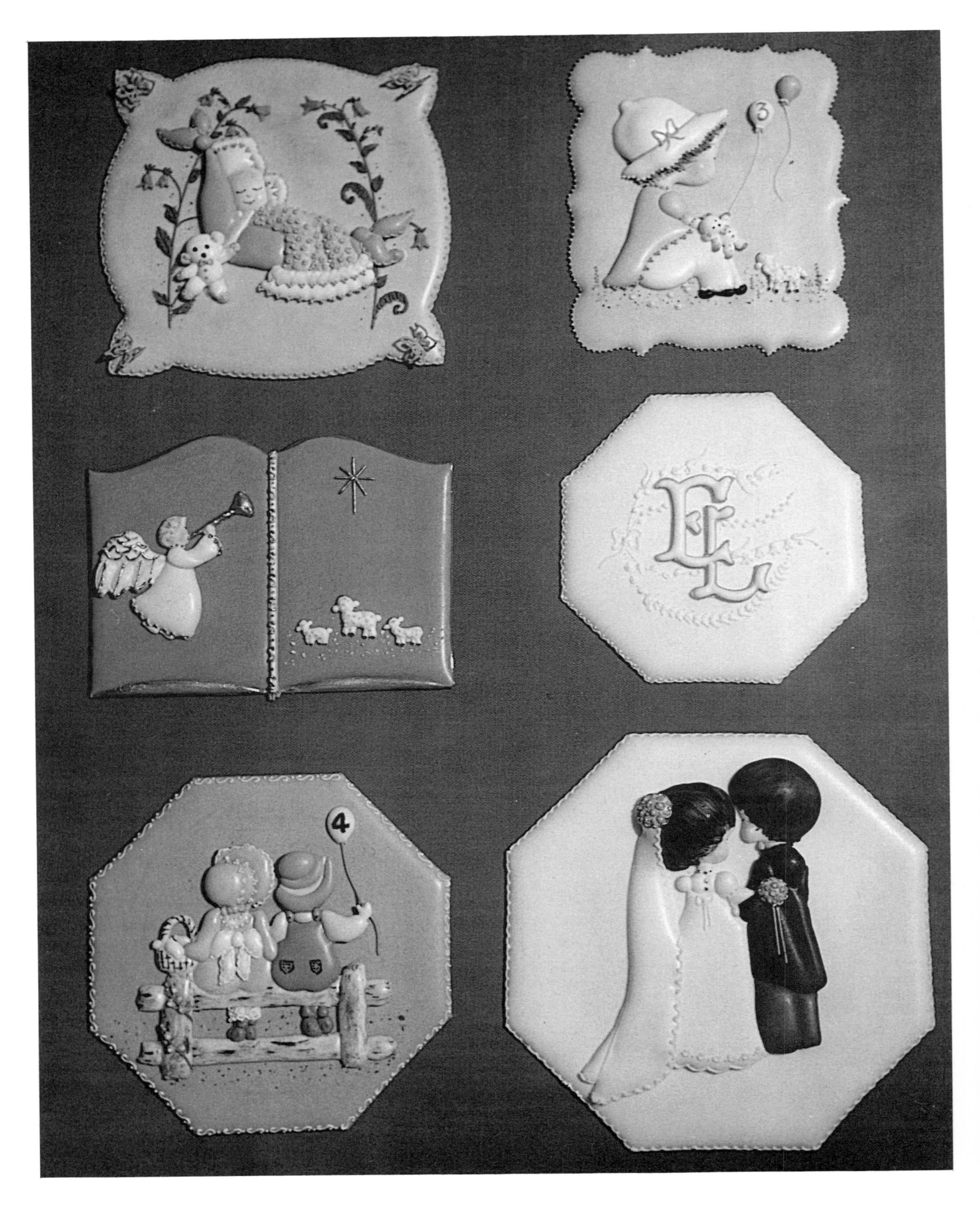

Fig. 73. Decorated plaques.

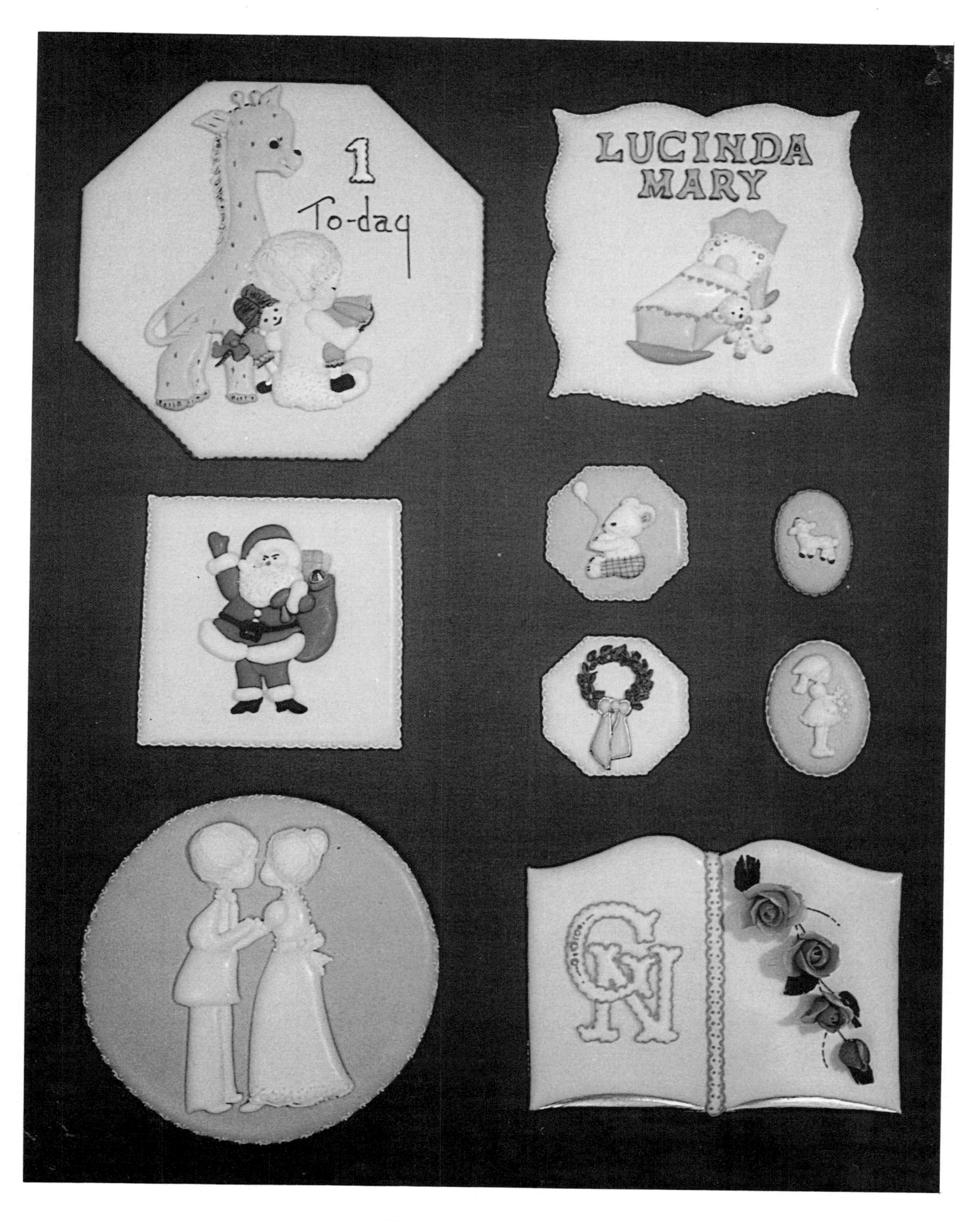

Fig. 74. Decorated plaques.

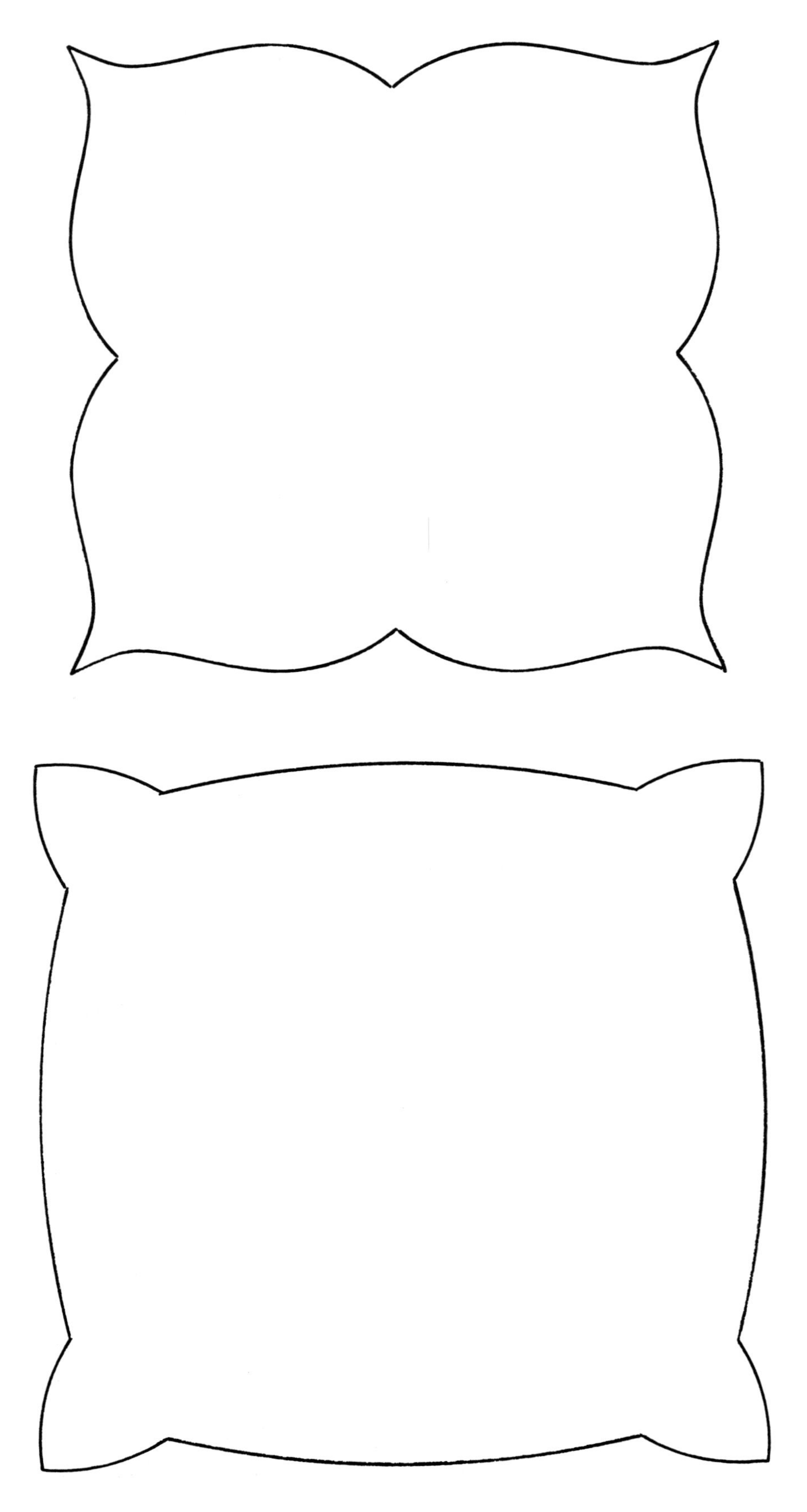

Fig. 75. Working drawings for plaques.

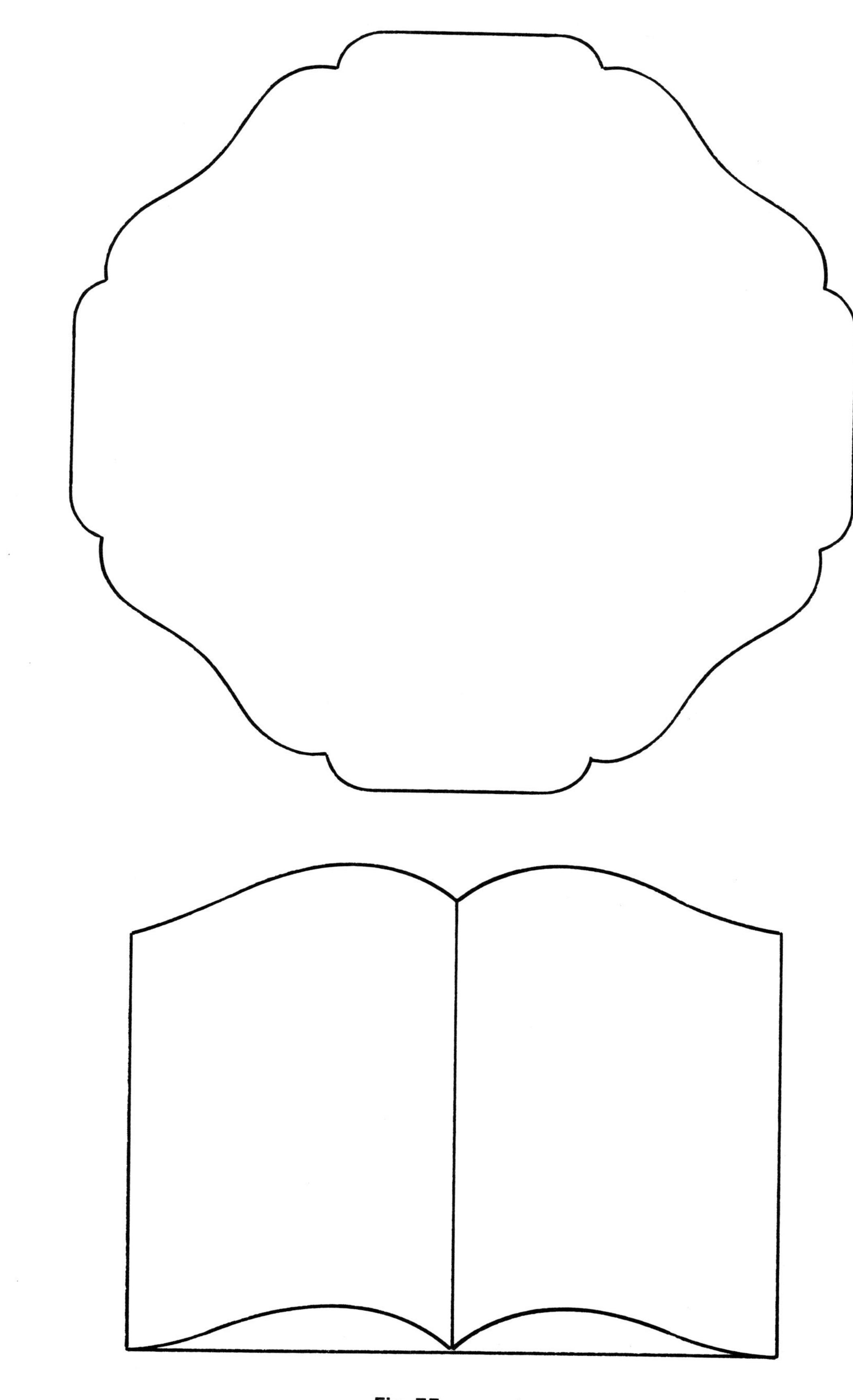

Fig. 75 — *contd.*

Fig. 75 — *contd.*

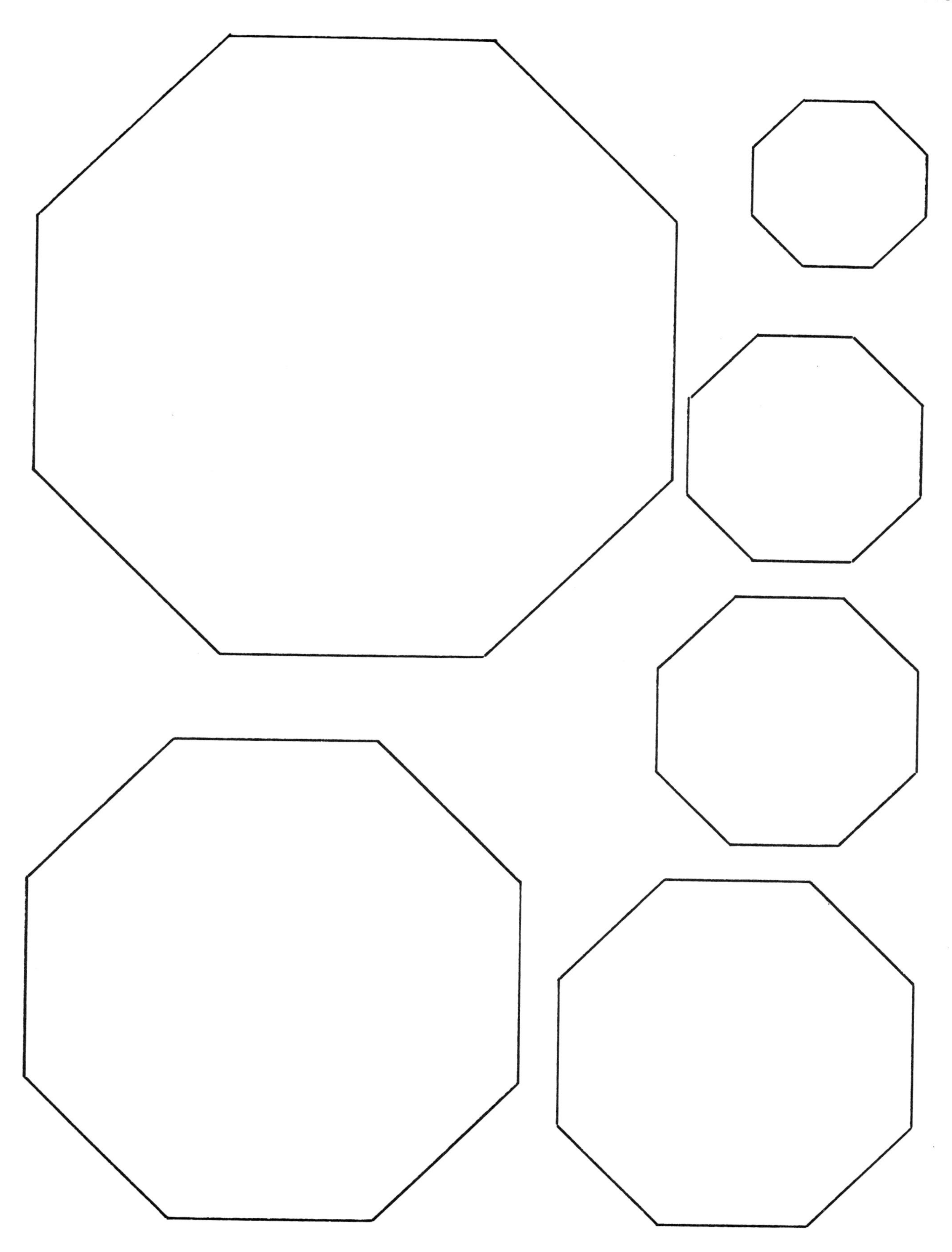

Fig. 75 — *contd.*

CHAPTER 11

Runout Figure Piping

For many reasons, runout figure piping is my favourite subject, but most of all because it expresses individuality.

Ideas for Drawing

Contrary to popular belief, it is not essential to be able to draw in order to do runout figure piping. All that is required is the ability to trace. Ideas come from many sources but mainly from gift-wrapping paper, children's painting books and greetings cards.

The cards illustrated in Fig. 76 are given as working drawings in Fig. 86. You will notice that each has, in some way, been adapted to suit our needs.

Fig. 76 Greetings cards.

Interpretation of Design

Experience plays a large part in being able to take a given design and make it suitable for a runout figure. Initially, a number of greetings cards will be examined and discarded as unsuitable. With practice however, unsuitable sections can be discarded and additional items included to form an acceptable design. The following are examples of this.

Father Christmas (*Fig. 76*)
The Father Christmas on this card has a sack which was not required although it could have been used, separately, for the side of the cake. In the working drawing (Fig. 86) a toy has been placed in the chimney.

Children on fence (*Fig. 76*)
This is an ideal drawing for several reasons. Firstly, faces are not visible and this is where the majority of runouts are spoiled. Secondly, although at first glance it would seem suitable only for a child's birthday cake, it could be used for a golden wedding or retirement cake with the inclusion of the appropriate numerals on the side of the cake or as part of the border design. Finally, the position of the boy's hand enables a balloon or a kite, depicting the age of the child, to be easily positioned when required for a birthday cake (working drawings Fig. 86).

This runout is shown on a cake in Plate 3 and on a plaque in Fig. 73.

Baby in hammock (*Fig. 76*)
This figure can be used with or without the tree; teddy-bear or rabbit. Hands are difficult to pipe and the left hand is not shown on the working drawing (Fig. 86). It can be indicated by a raised

Plate 1. Christmas cake — 18 cm (7 in.) cake on 28 cm (11 in.) board.

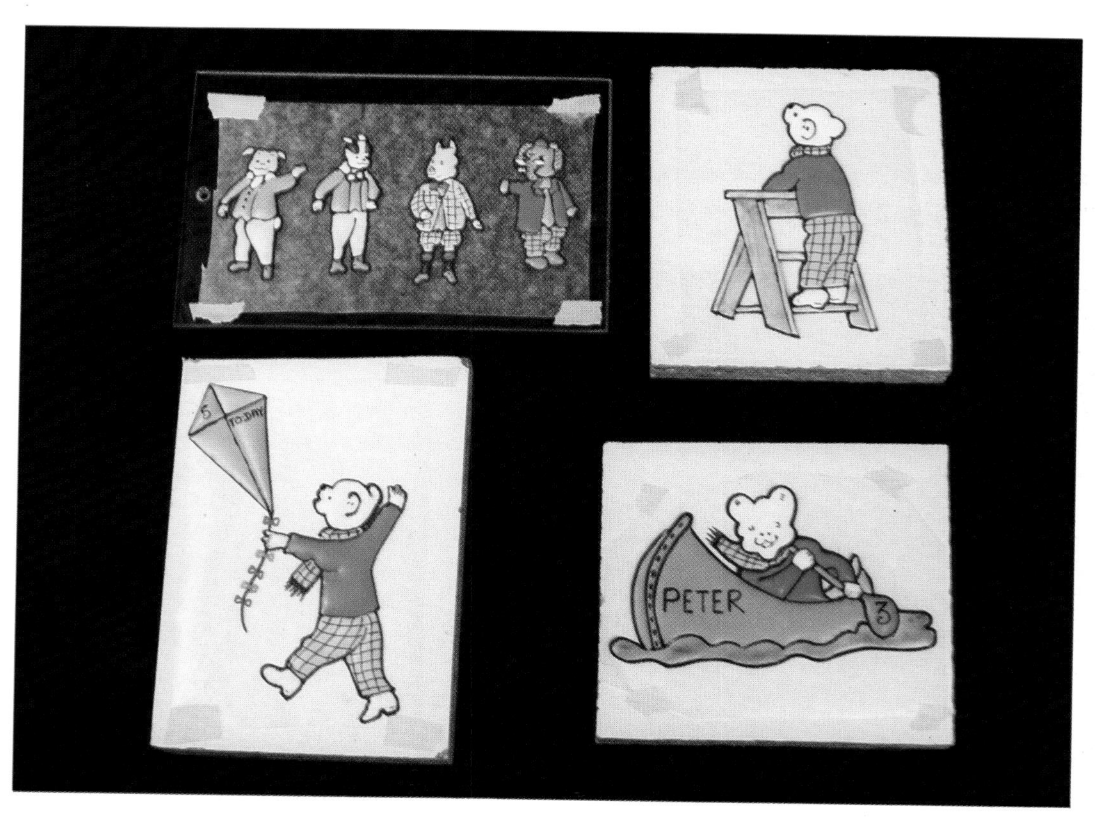

Plate 2. Outlined figures.

Plate 3. Birthday cake — 15 cm (6 in.) cake on 23 cm (9 in.) board.

Plate 4. Engagement cake — 20 cm (8 in.) cake on 30 cm (12 in.) board.

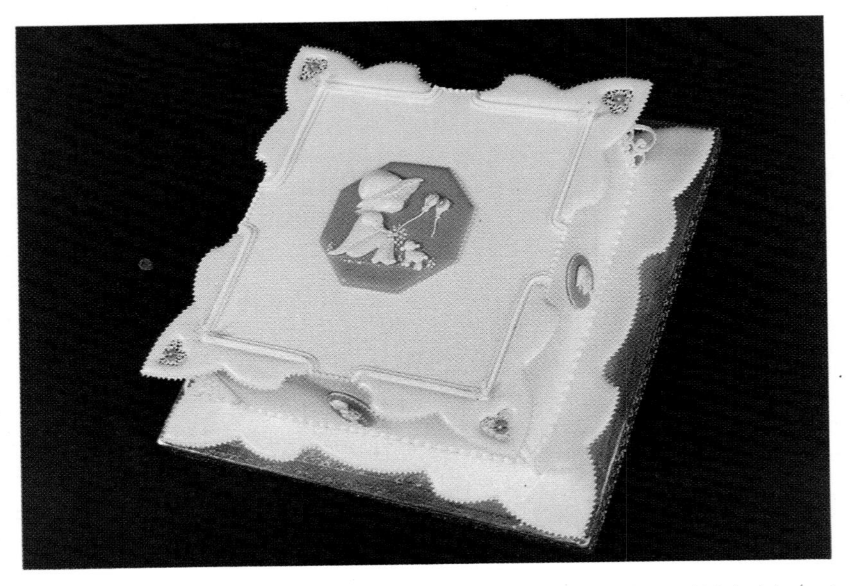

Plate 5. 70th birthday cake — 20 cm (8 in.) cake on 30 cm (12 in.) board.

Plate 6. Completed figures.

Plate 7. Father Christmas — step-by-step.

section under the blanket. The teddy-bear could be runout separately and placed on the figure later.

This runout is shown on a cake in Fig. 50 and on a plaque in Fig. 73.

Engaged couple (*Plate 4*)
This working drawing (Fig. 86) can be made suitable for a wedding anniversary by the addition of a veil or bouquet. By the same rule a veil can be omitted from a drawing to make it suitable for some other occasion.

Realism

A common fault in translating a given design into one suitable for runout work is to endeavour to make it too realistic. Of course, a certain amount of realism is required and this is why I prefer this type of figure work to others, but we must never forget that an edible decoration is being produced. An example of this is the mouse under the tree (Fig. 82): it is acceptable as an edible decoration because it is not portrayed in a realistic way and the background piping is unrealistic also. The illustration is shown on a plaque for the purpose of the exercise, but generally the design is more effective if placed directly on a cake.

Use of Colour

When deciding on a design for a runout it is necessary to decide upon the colour or colours to be used. Most figures require bright colours and whilst this is acceptable for children's cakes, it is not suitable for every occasion. When brightly coloured figures are used, the coating and border work should be kept very pale.

Flesh Colour

Flesh colour is obtained by mixing a small amount of pink with an even smaller amount of yellow. Paste colours such as 'coppertone' and 'paprika' also produce a colour suitable for faces. It must be emphasized that the icing should appear off-white when sufficient colour has been added to ensure that it will dry to the correct shade. The only exception to this is a Father Christmas where a ruddier complexion is required.

Blue Colour

The use of blue colouring deserves a mention. Personally I do not usually coat a cake with blue icing, because blue fades very quickly. If, however, they are kept covered, blue plaques will not fade and they can be used where a blue cake is required. White figures on blue look most attractive and give a Wedgwood effect. The figures and plaques can be runout in advance and the plaque assembled on a white cake shortly before it is required. I recommend that should a blue cake be required, this form of decoration is used (Plate 5).

Piping/Flooding

The secret of successful runout figure piping is the consistency of the icing. It is almost impossible to describe; it has to be seen and, preferably, felt. For this reason I have included step-by-step photographs depicting the piping of several runout figures (Figs 77 to 85). I should point out, however, that apart from certain prominent features, the order of work is a personal matter provided an area is not piped before the adjoining one has crusted.

Consistency

The consistency required for runout figures is thicker than that used for other runouts. The icing is reduced to a thick cream which does not flow but will lose marks made by a paint brush. Some people advise counting to see how many seconds it takes for the brush marks to disappear but I find this method confusing and inaccurate.

Most students find the correct consistency easy to achieve after very little practice. Start by letting down the icing *very slightly*. Try it on a board, using a paint brush to move the icing gently. If this is too thick, add a little more egg white/albumen and so forth until the icing will stay where required without leaving the mark of the paint brush.

A few people obtain the right consistency but still have difficulty in obtaining a smooth finish to the runout figure. This is usually because they have taken too long to flood the area. It is a *quick* action, which may call for the use of the brush momentarily to adjust the position of the icing. Spending too much time will result in thickening of the icing with persistence of brush marks.

Some areas will require to be flooded thicker than others and this is achieved by pressure on the bag. If the consistency is correct it is possible to flood the area as thickly as desired whilst still achieving a smooth finish (Fig. 77).

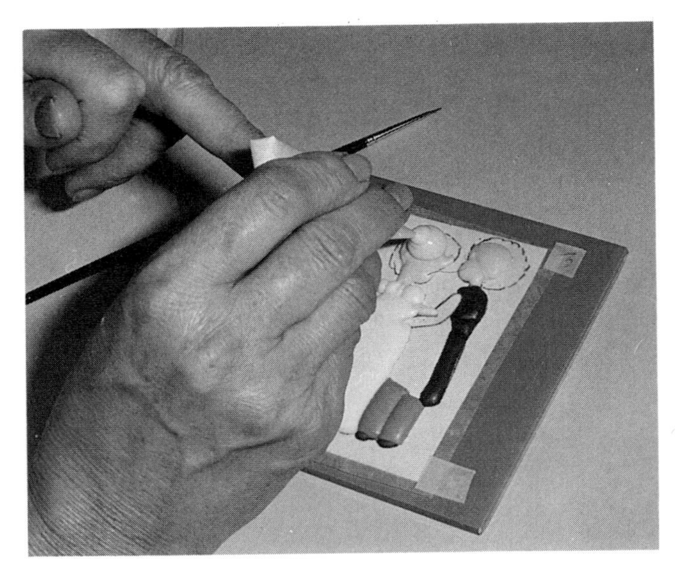

Fig. 77 Hair on runout figure.

Runout figures can be left unfinished provided the same colour of icing is available for completion and has been reduced to the correct consistency.

Points to Remember for Runout Figure Piping

1. The basic principles of runout work are outlined in chapter 8 and apply to runout figure piping.
2. Small bags of icing are used and the hole cut should not be larger than a No. 2 tube.
3. Piping tubes are not required.
4. A more realistic effect is achieved when icing is flooded over, rather than to, an outline – as when piping faces. For this method it is essential that a spare drawing is used as a guide to piping.
5. To avoid contamination of one flooded area by a neighbouring one, strong colours should remain under a lamp for a longer period than pale colours.
6. Make sure that every colour required is available in small bags before the work is commenced. Each bag should not be more than half full and should indicate the colour of icing it contains.
7. Bags of icing should not be held when not in use as heat from your hand will alter the consistency.
8. Small amounts of icing are required for figures. Colour is added with a cocktail stick and 'paddled down' on a work surface with a small palette knife before adjusting the consistency with egg white or albumen.
9. Icing in bags will stay the correct consistency (if not unduly handled) for approximately two hours. If it should become too thick, a larger hole is cut in the bag; the icing squeezed out, adjusted with egg white or albumen and placed in a fresh bag.
10. Icing that has become too thick for flooding may be suitable when a rough texture is required – eg. trees, snow, Father Christmas's beard.

Finishing Touches

These are very important and unless skilfully done can spoil a figure that has been satisfactorily flooded (Fig. 78 and Plate 6).

Fig. 78 Father Christmas runout and assembled.

Painting
All painting is carried out after the figures are dry and whilst still attached to the waxed paper. A very fine sable hair paint brush should be kept for this purpose.

Full faces
Wherever possible choose a design with a profile or back view. If a full face is required it should be painted very carefully using an almost dry paint brush and edible paste colours. The features should be traced and gently transferred to the figure with a pencil before painting. Do not *guess* where the eyes should be, it will almost certainly be wrong! Exceptions to this are Father Christmas or small animals (lambs etc.) where a realistic result is not required.

Profile
A small mark of black colouring where the eye should be is all that is required.

Cheeks
Unless a figure is to be entirely white, cheeks will require a *little* colour. This is done by brushing with a little 'blossom tint' or 'petal dust' (skintone) which has been toned down with cornflour. The brush must be dry and the powder used very sparingly.

Hair
Hair is runout a paler shade than required and then 'streaked' with the correct colour using a fine paint brush. Blonde hair is achieved by using pale yellow for the flooding and when dry, streaking with a mixture of yellow, orange and a little

(*text continued p. 168*)

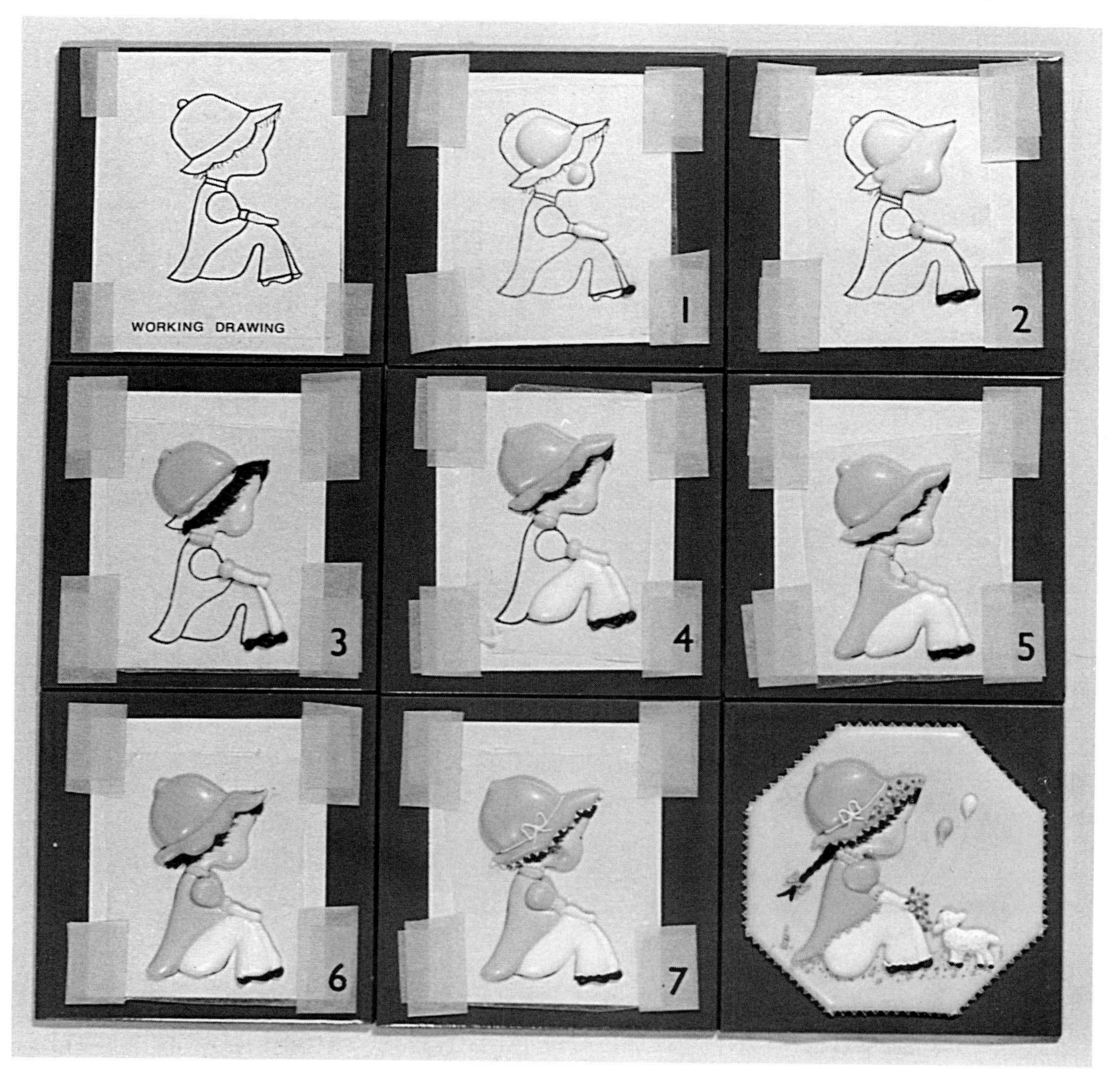

Fig. 79 Sitting child — step-by-step.

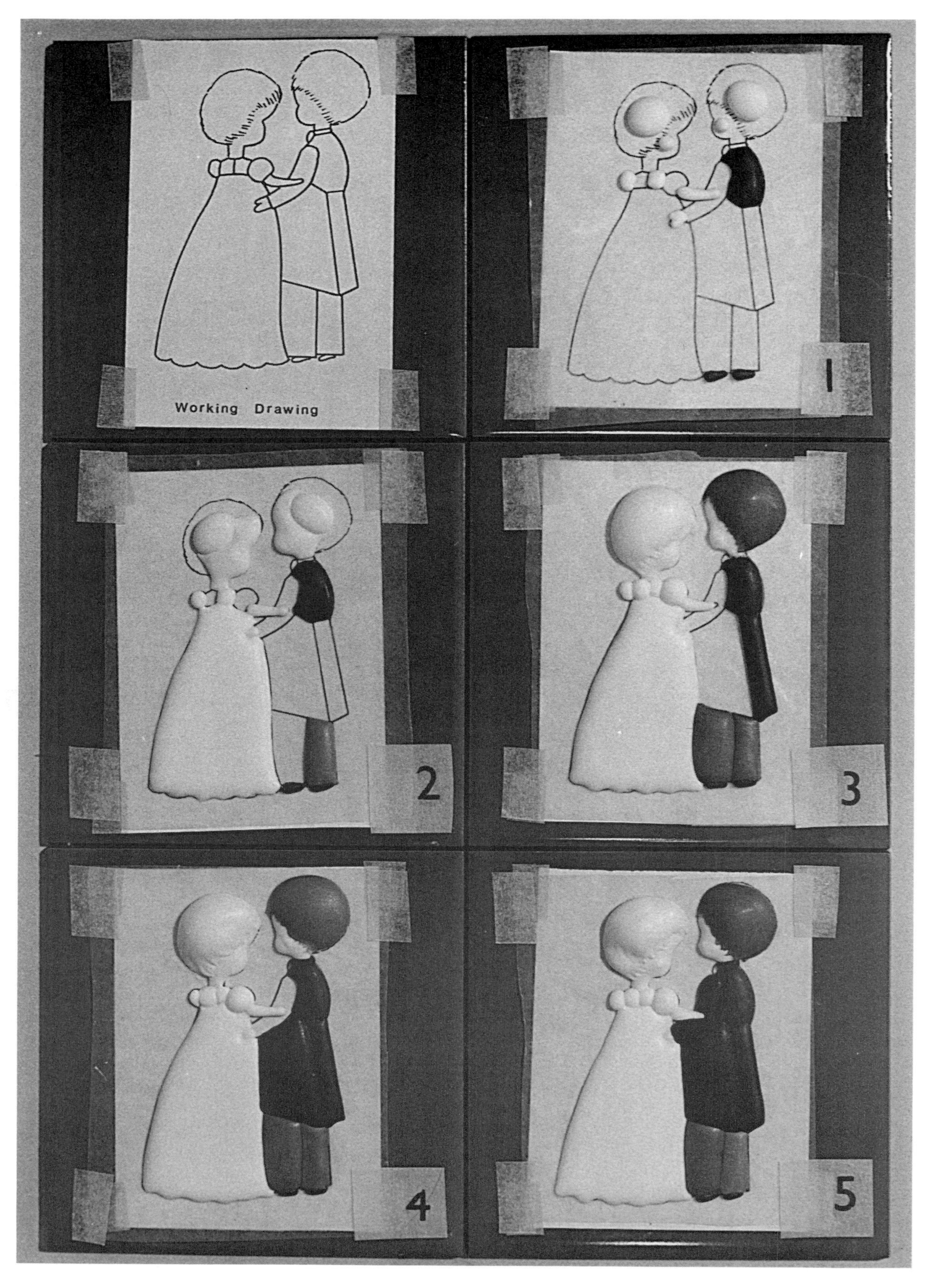

Fig. 80 Bride and groom — step-by-step.

Fig. 80 — *contd.*

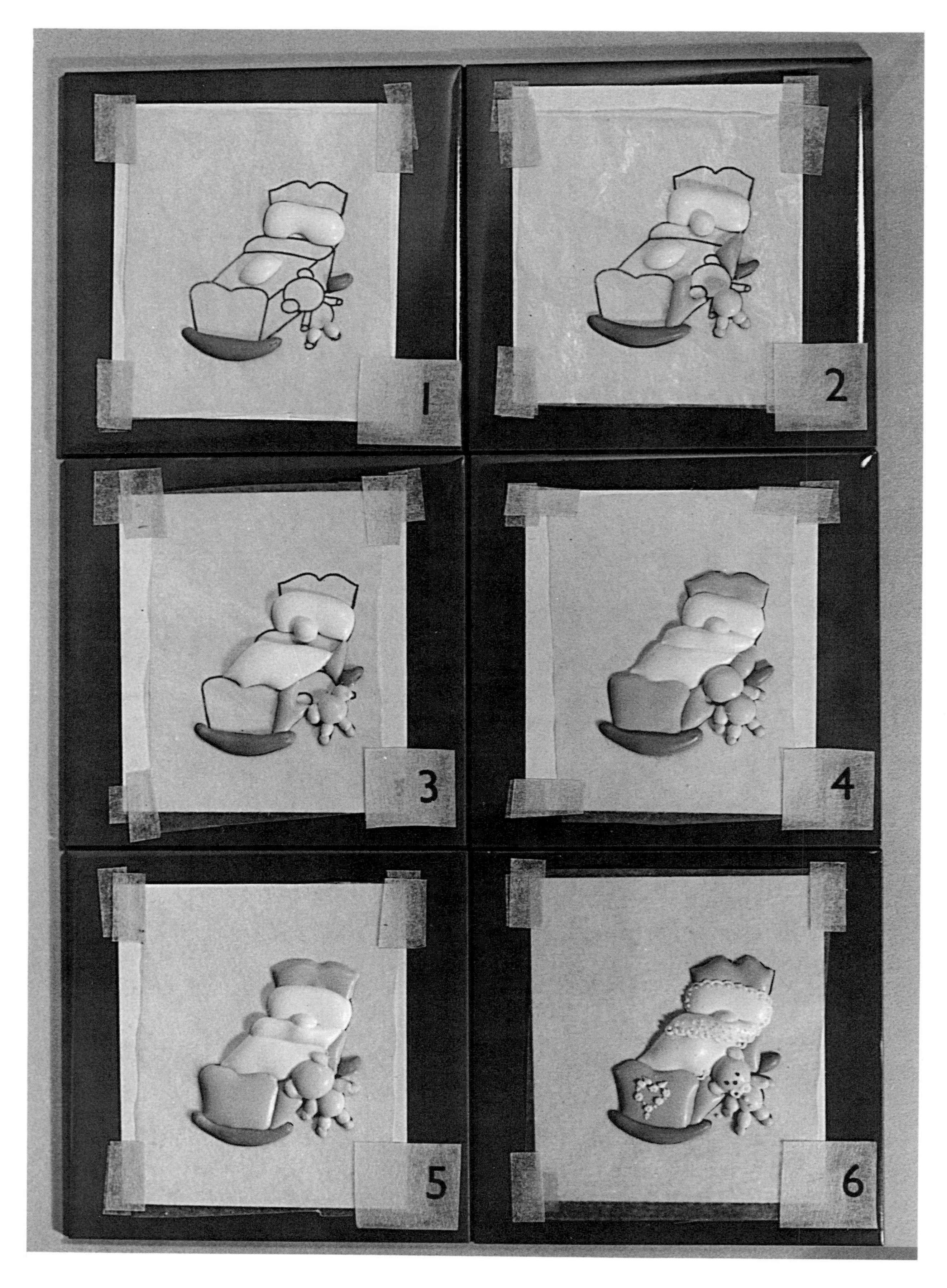

Fig. 81 Cradle — step-by-step.

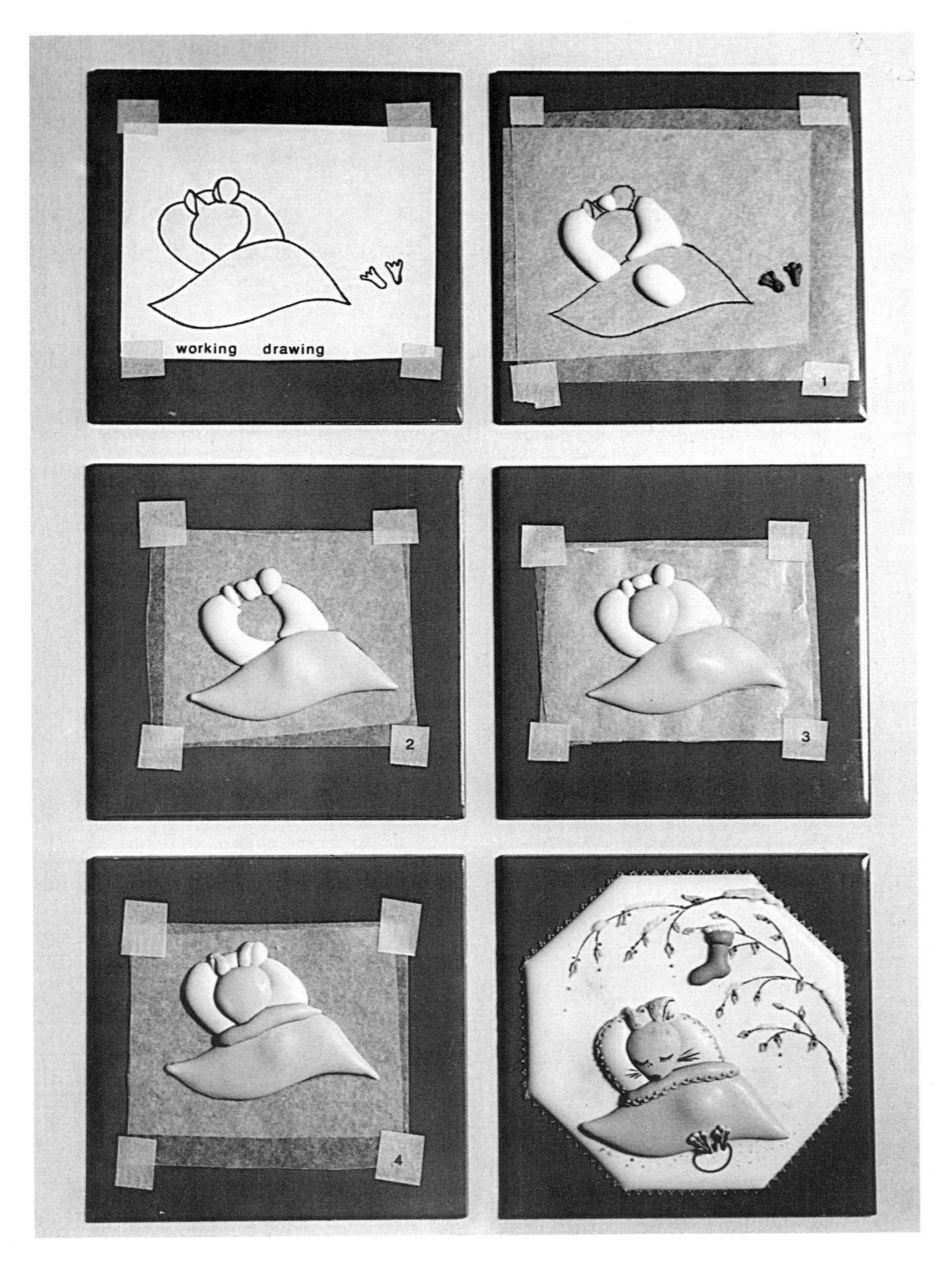

Fig. 82 Mouse — step-by-step.

Fig. 83 Girl with lambs — step-by-step.

Fig. 84 Girl with lambs on plaque.

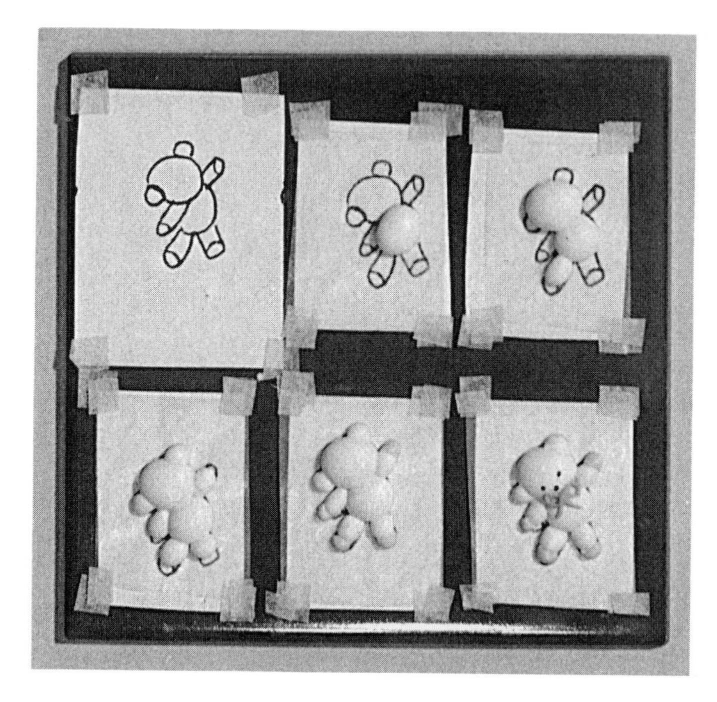

Fig. 85 Lambs and bears — step-by-step.

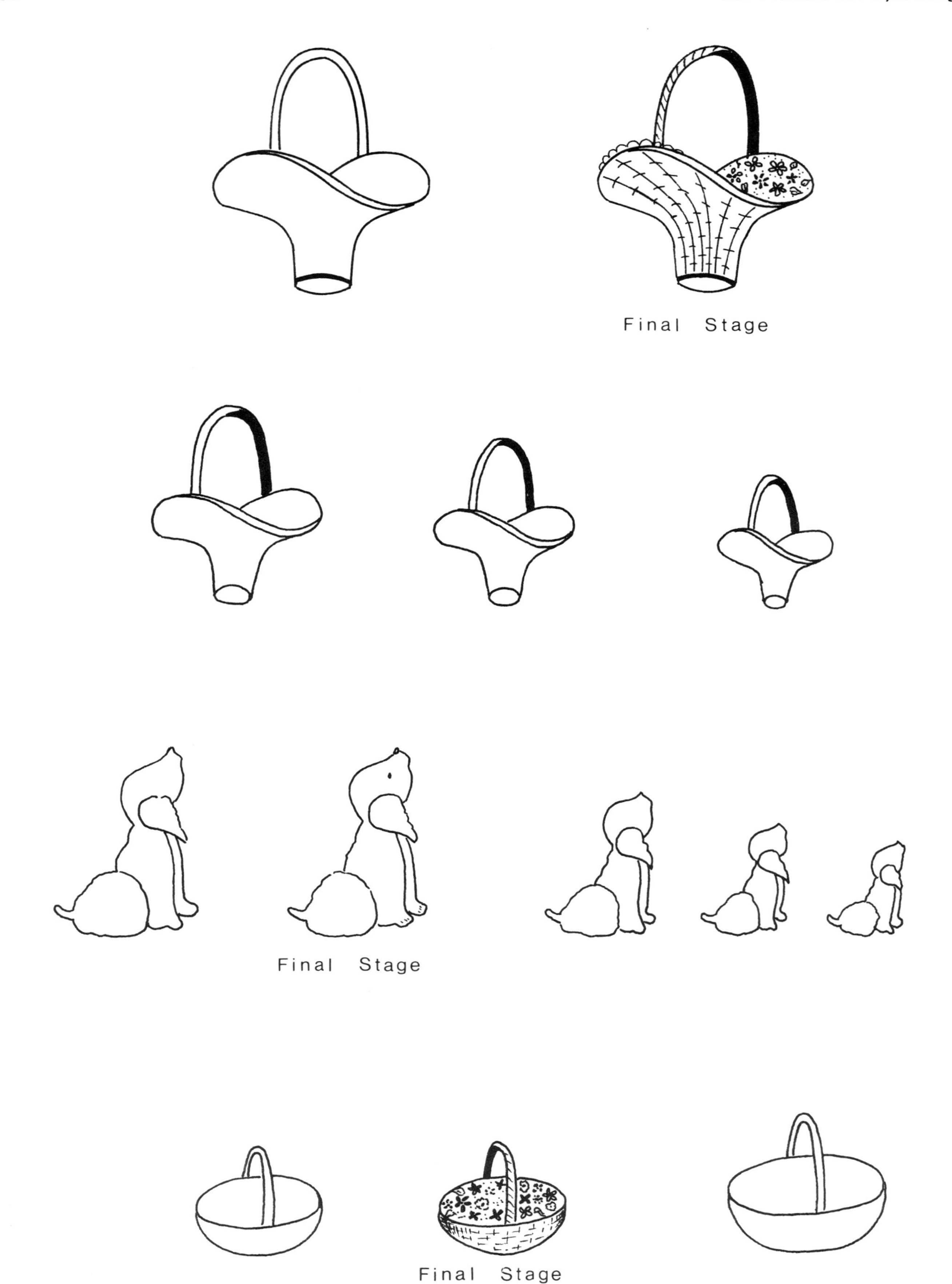

Fig. 86 Working drawings for runout figure piping.

2
Working Drawings
Intermediate Stage
Final Stage
Working Drawings
Final Stage

Fig. 86 — *contd.*

Fig. 86 — *contd.*

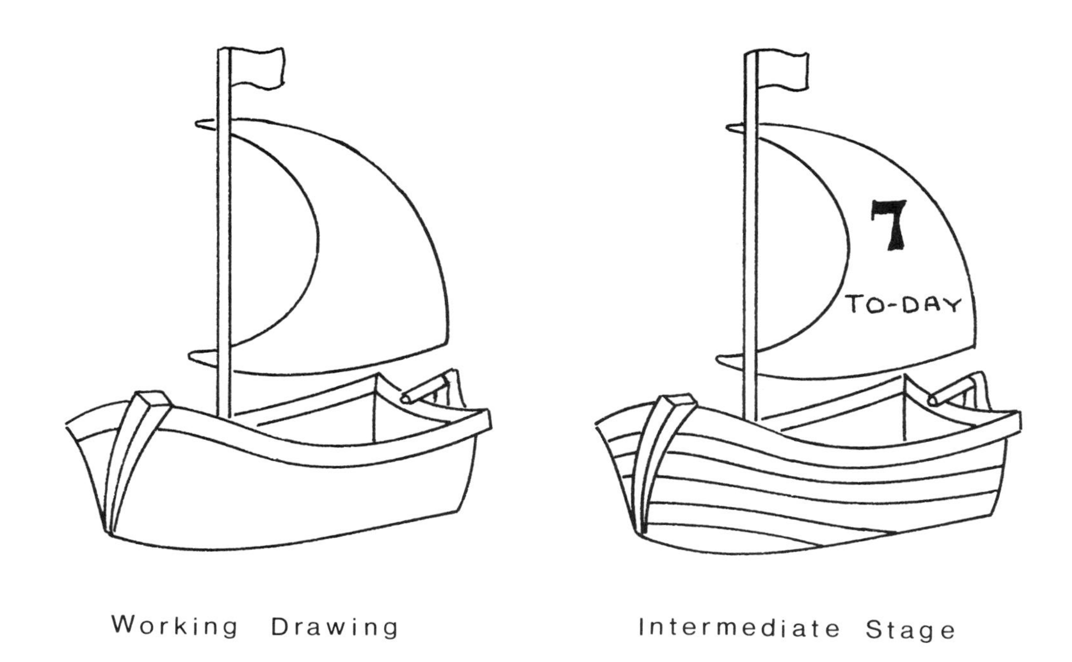

Working Drawing Intermediate Stage

Final Stage

Fig. 86 — *contd.*

Fig. 86 — *contd.*

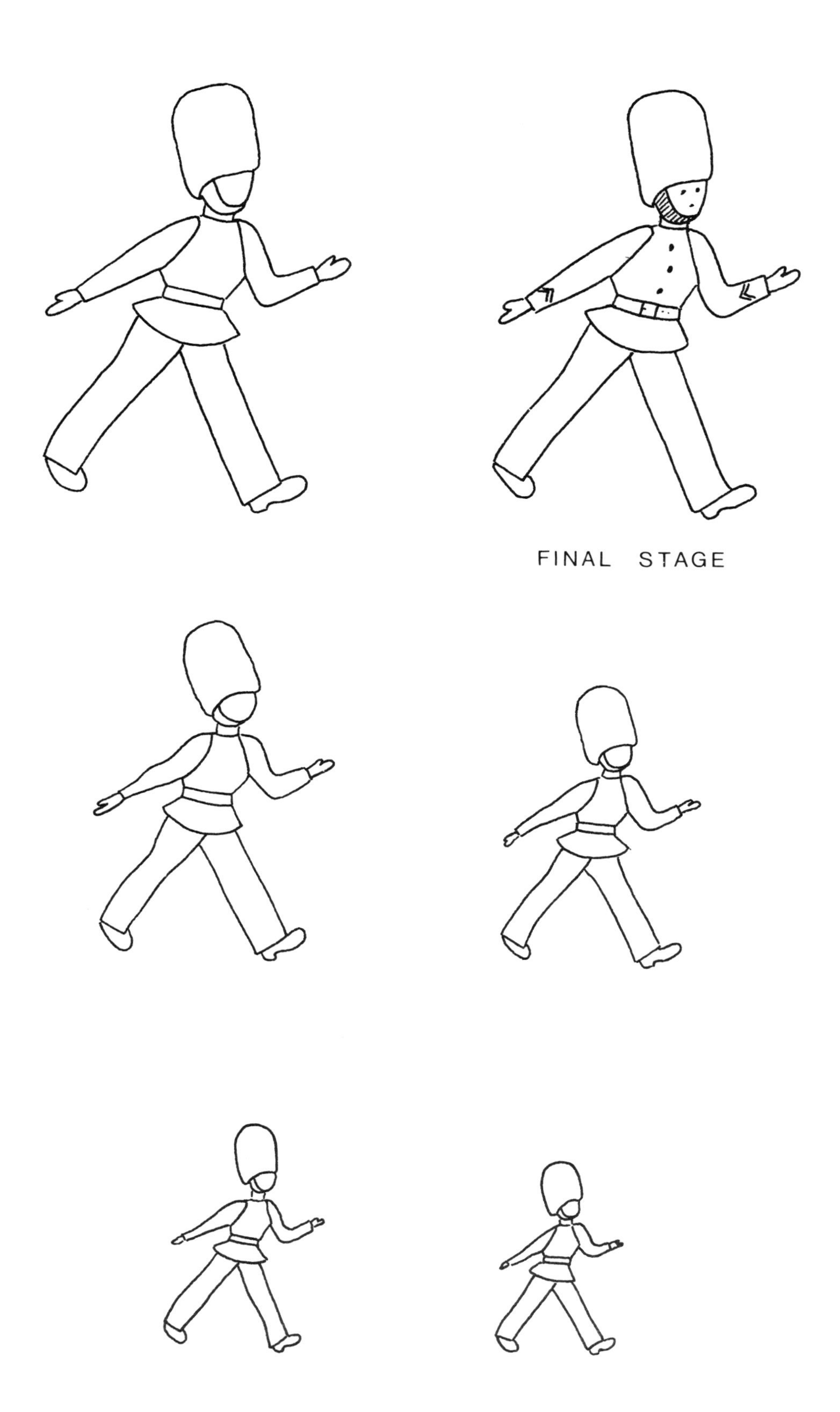

Fig. 86 — *contd.*

Fig. 86 — *contd.*

FINAL STAGE

Fig. 86 — *contd.*

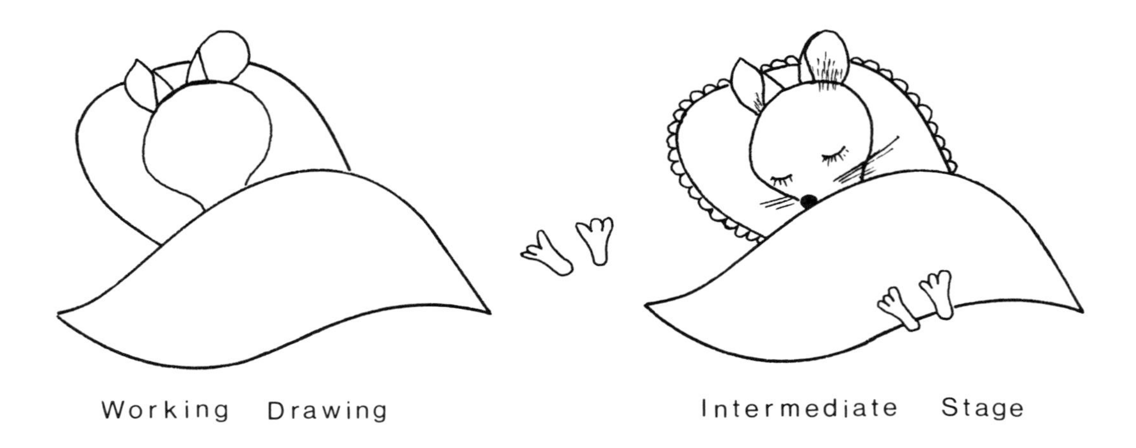
Working Drawing
Intermediate Stage

Final Stage

Working Drawings

Fig. 86 — *contd.*

Fig. 86 — *contd.*

Fig. 86 — *contd.*

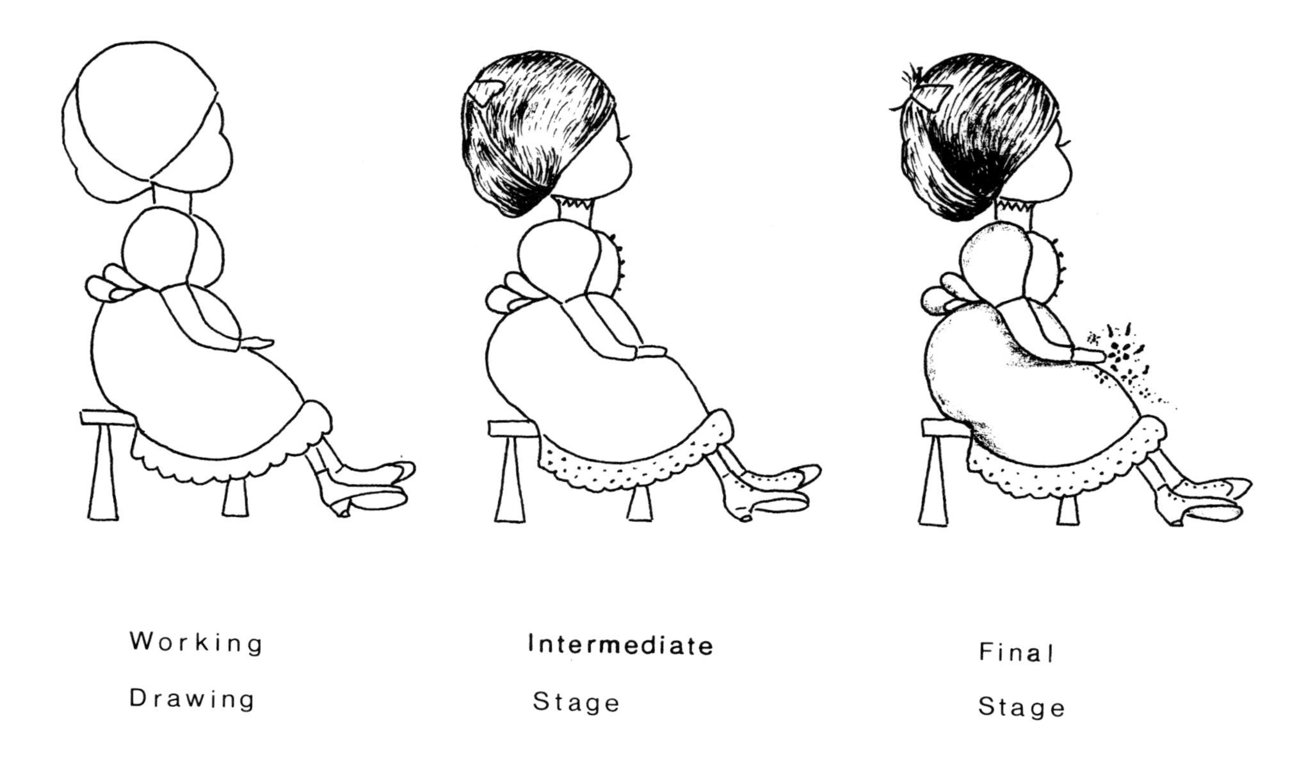
Working
Drawing
Intermediate
Stage
Final
Stage

Working Drawings

Fig. 86 — *contd.*

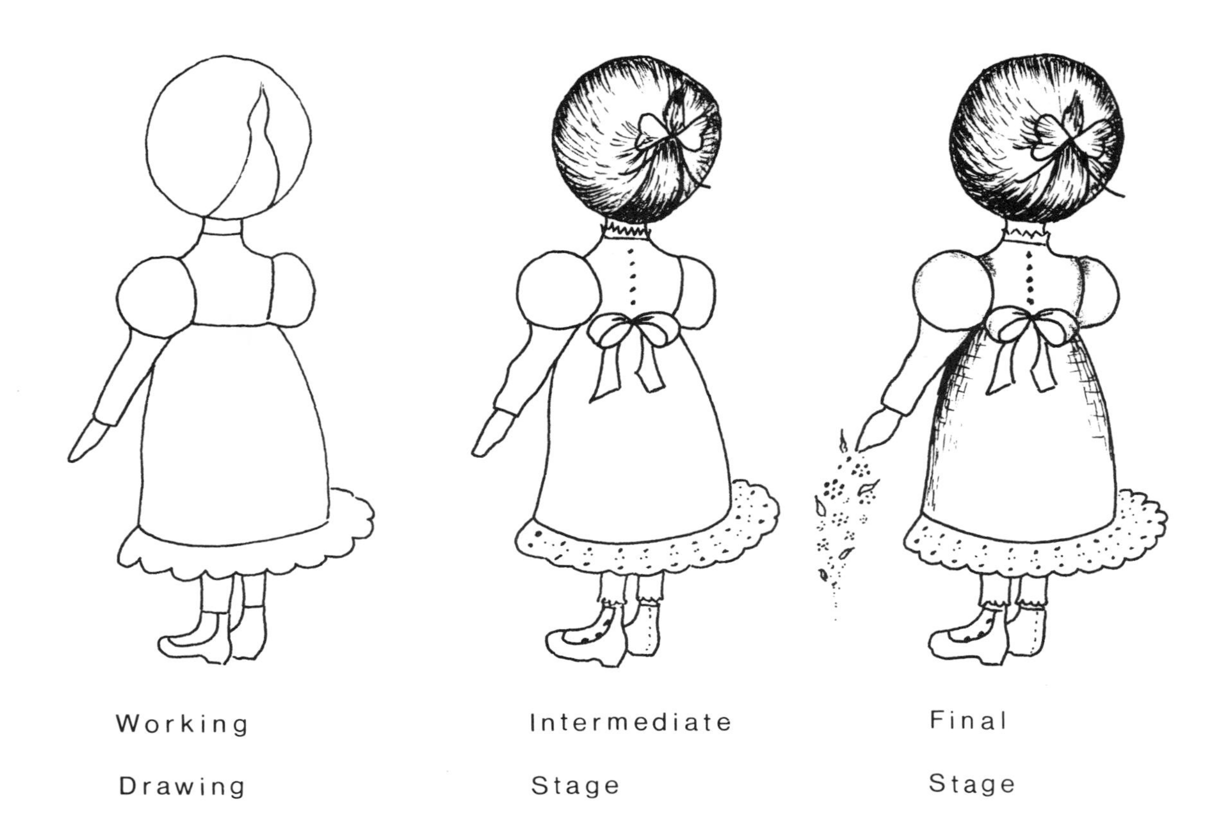
Working
Drawing
Intermediate
Stage
Final
Stage

Working Drawings

Fig. 86 — *contd.*

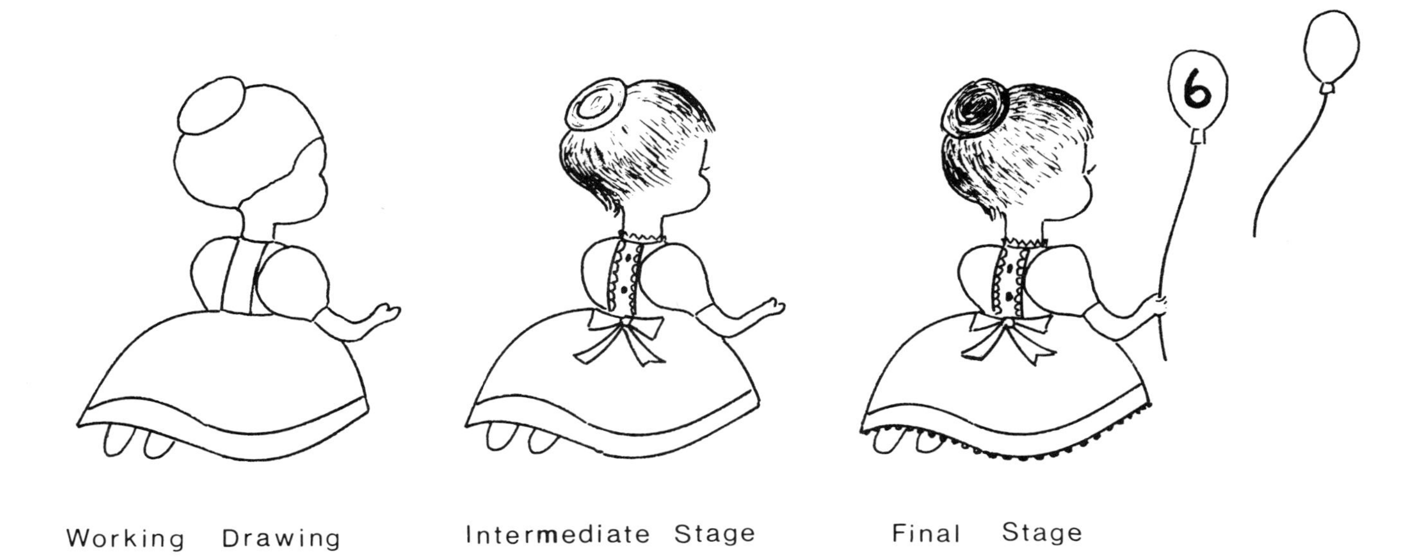
6
Working Drawing
Intermediate Stage
Final Stage

Working Drawings

Fig. 86 — *contd.*

Working Drawing

Intermediate Stage

Final Stage

Working Drawings

Fig. 86 — *contd.*

Fig. 86 — *contd.*

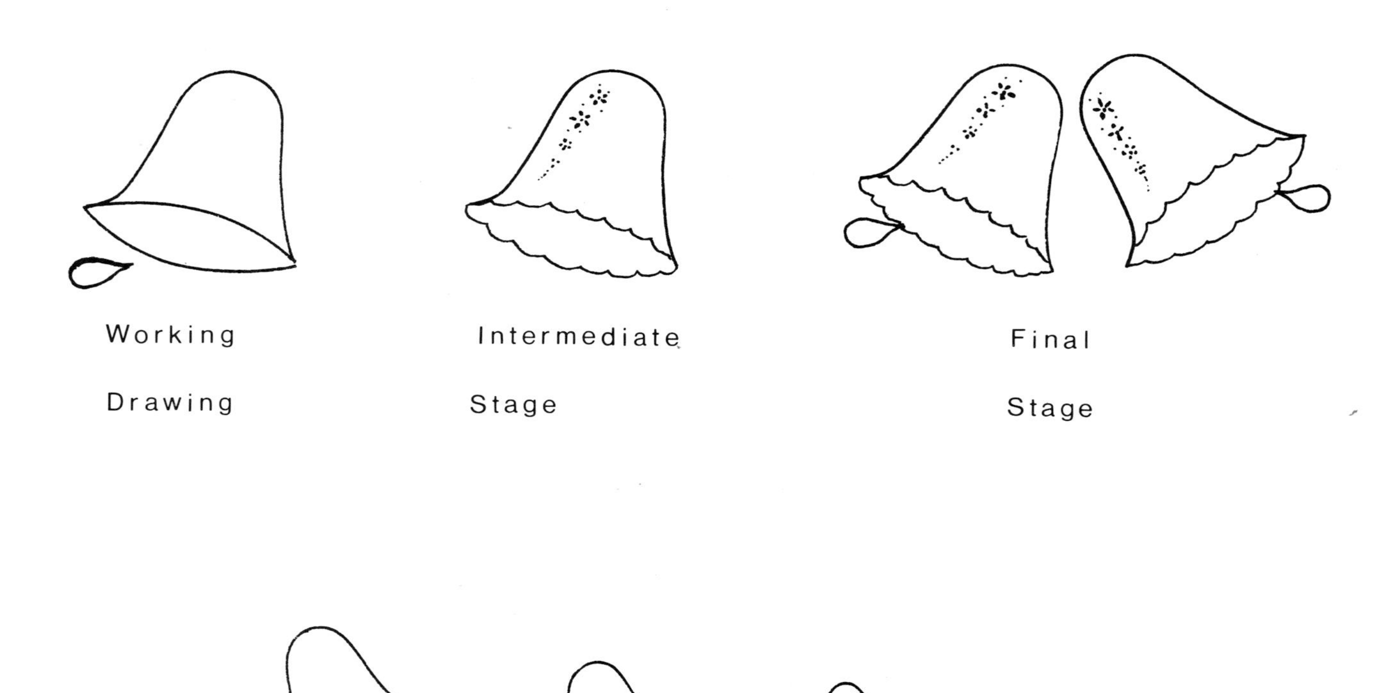

WORKING DRAWINGS

Fig. 86 — *contd.*

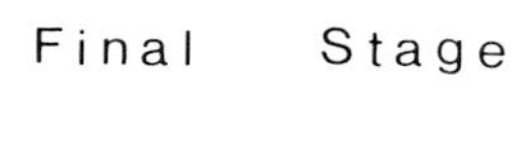

Working Drawings

Fig. 86 — *contd.*

Fig. 86 — *contd.*

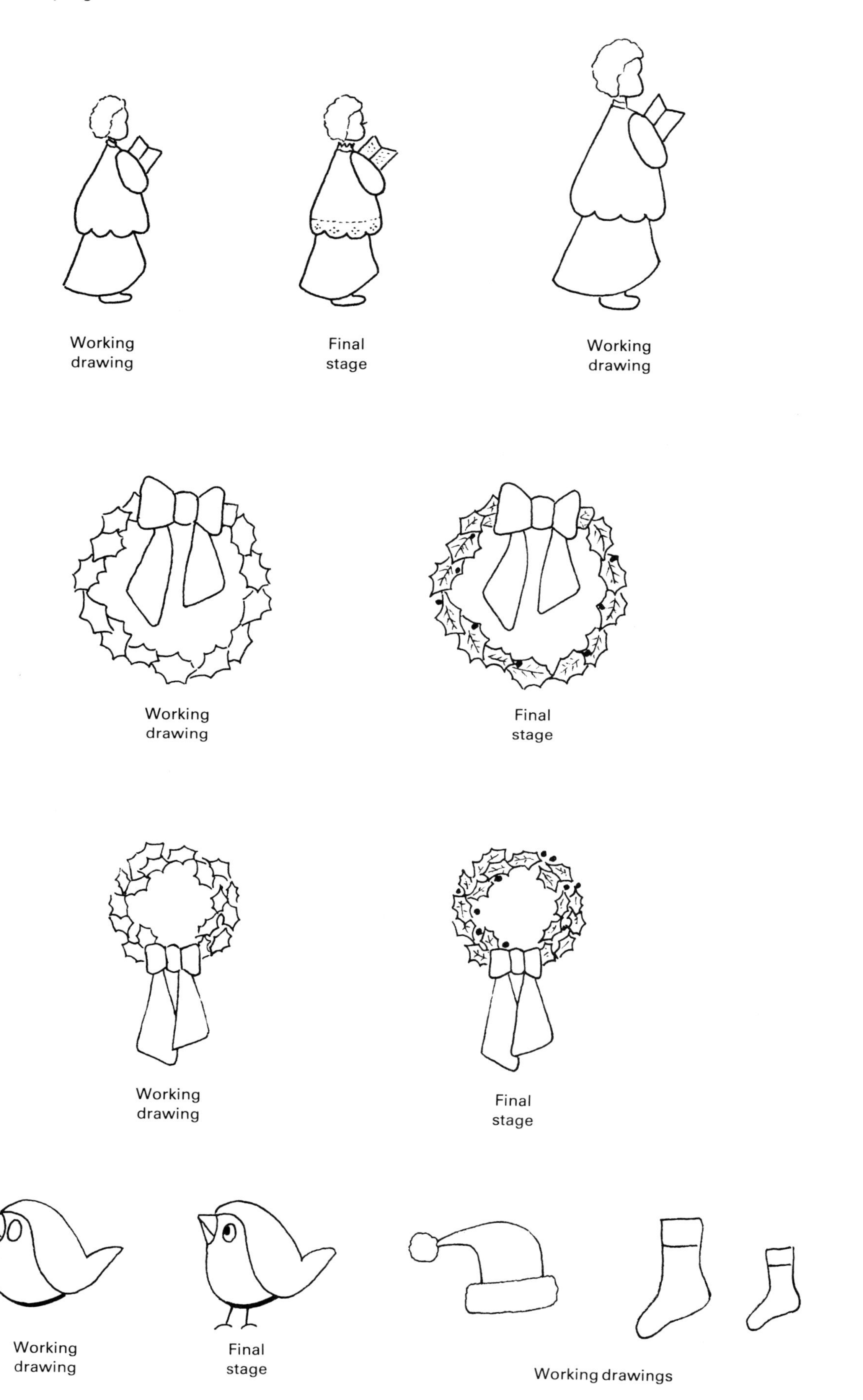
Working
drawing
Final
stage
Working
drawing
Working
drawing
Final
stage
Working
drawing
Final
stage
Working
drawing
Final
stage
Working drawings

Fig. 86 — *contd.*

Fig. 86 — *contd.*

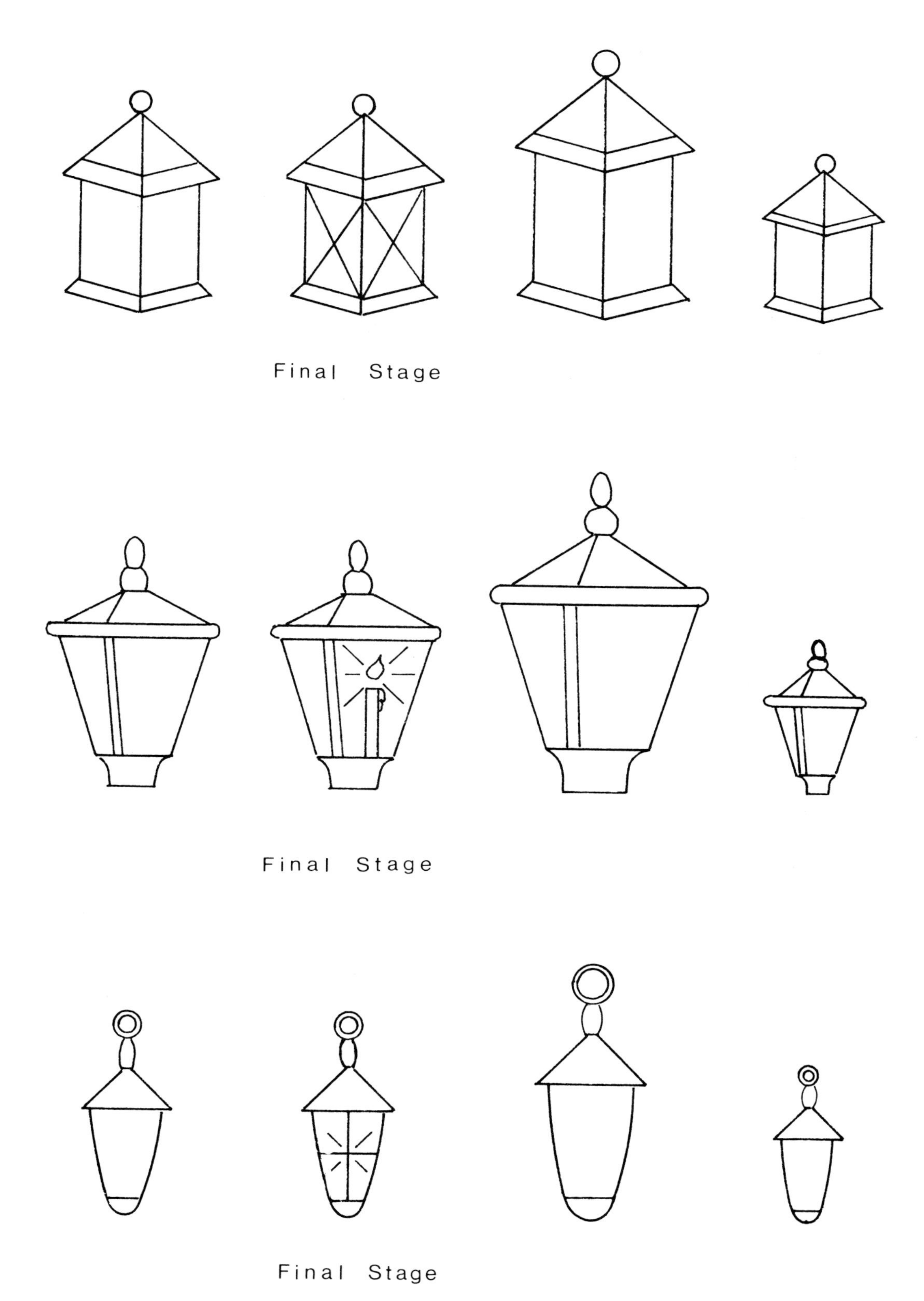

Fig. 86 — *contd.*

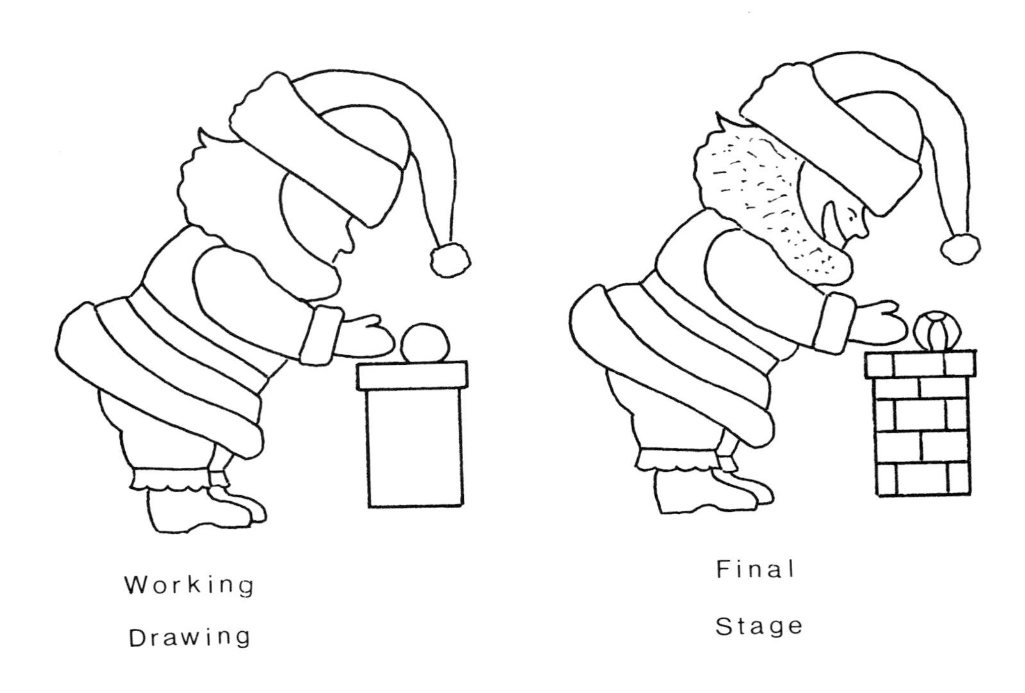

Fig. 86 — *contd.*

Final Stage

Fig. 86 — *contd.*

FINAL STAGE

Fig. 86 — *contd.*

WORKING DRAWINGS

FINAL . STAGE

Fig. 86 — *contd.*

Fig. 86 — *contd.*

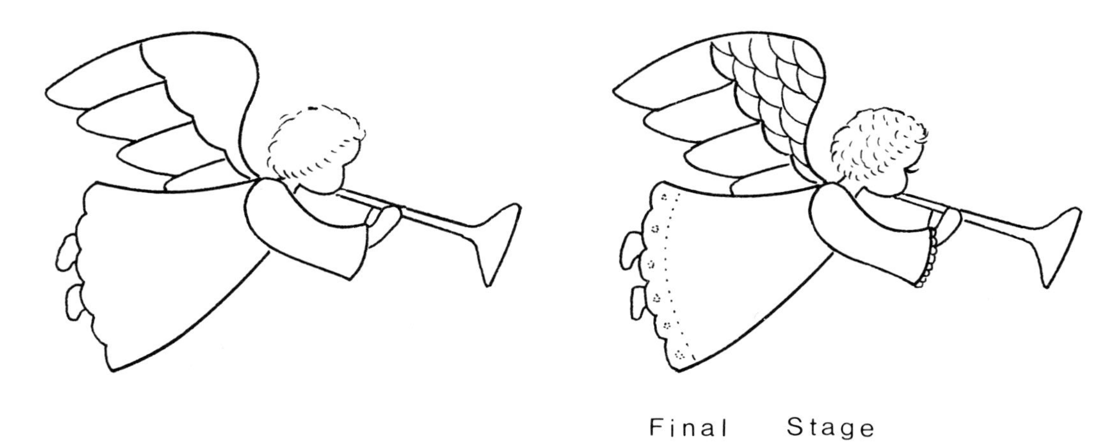

WORKING DRAWINGS

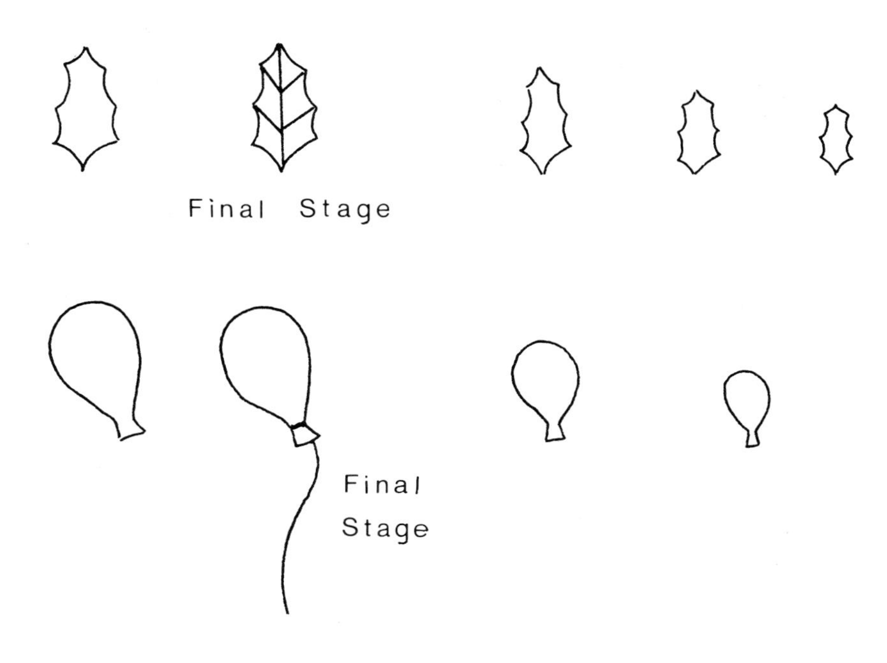

Fig. 86 — *contd.*

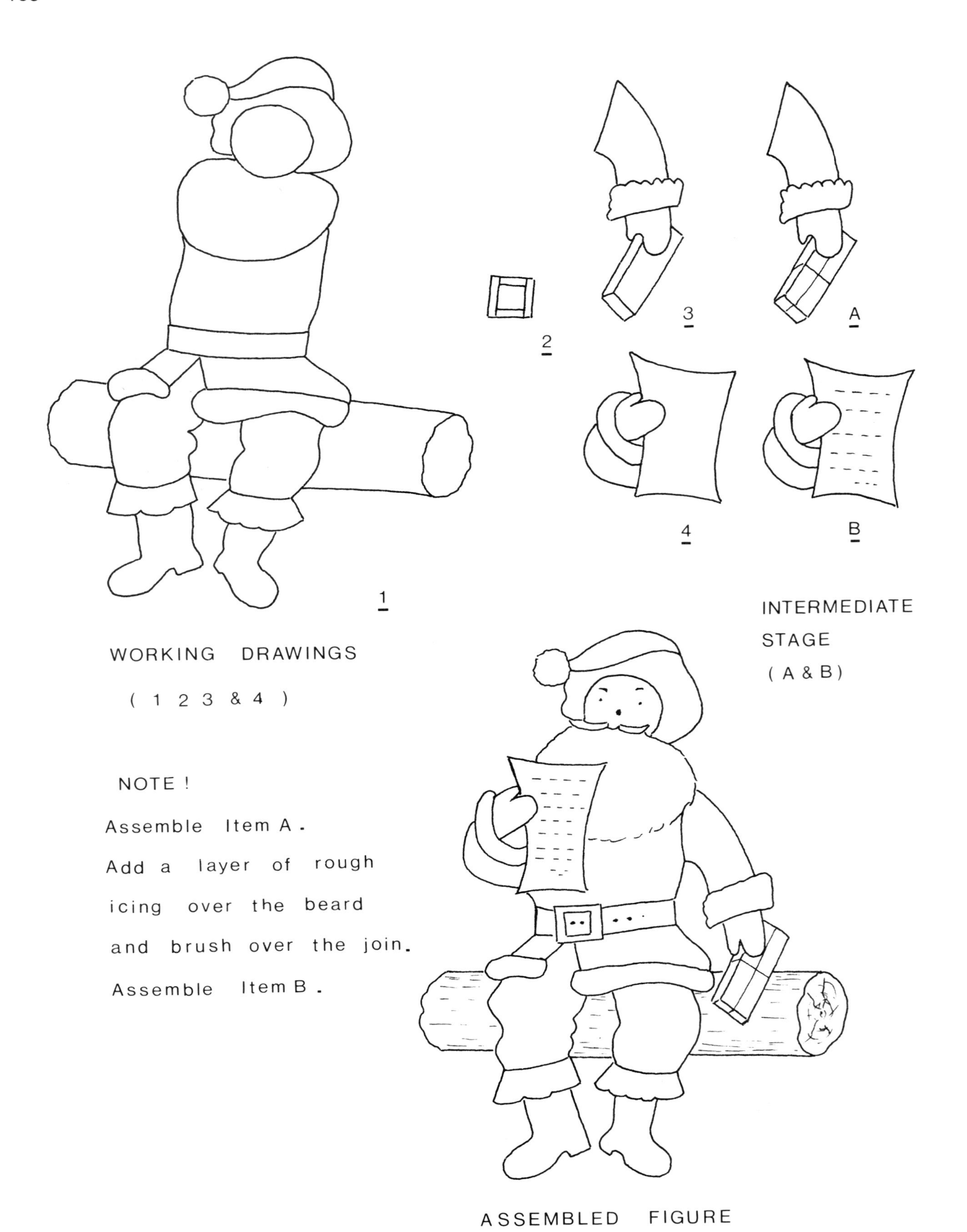
1
2
3
A
4
B
WORKING DRAWINGS
(1 2 3 & 4)
INTERMEDIATE
STAGE
(A & B)
NOTE !
Assemble Item A.
Add a layer of rough
icing over the beard
and brush over the join.
Assemble Item B.
ASSEMBLED FIGURE

Fig. 86 — *contd.*

WORKING DRAWINGS

Fig. 86 — *contd.*

Fig. 86 — *contd.*

WORKING DRAWINGS

Fig. 86 — *contd.*

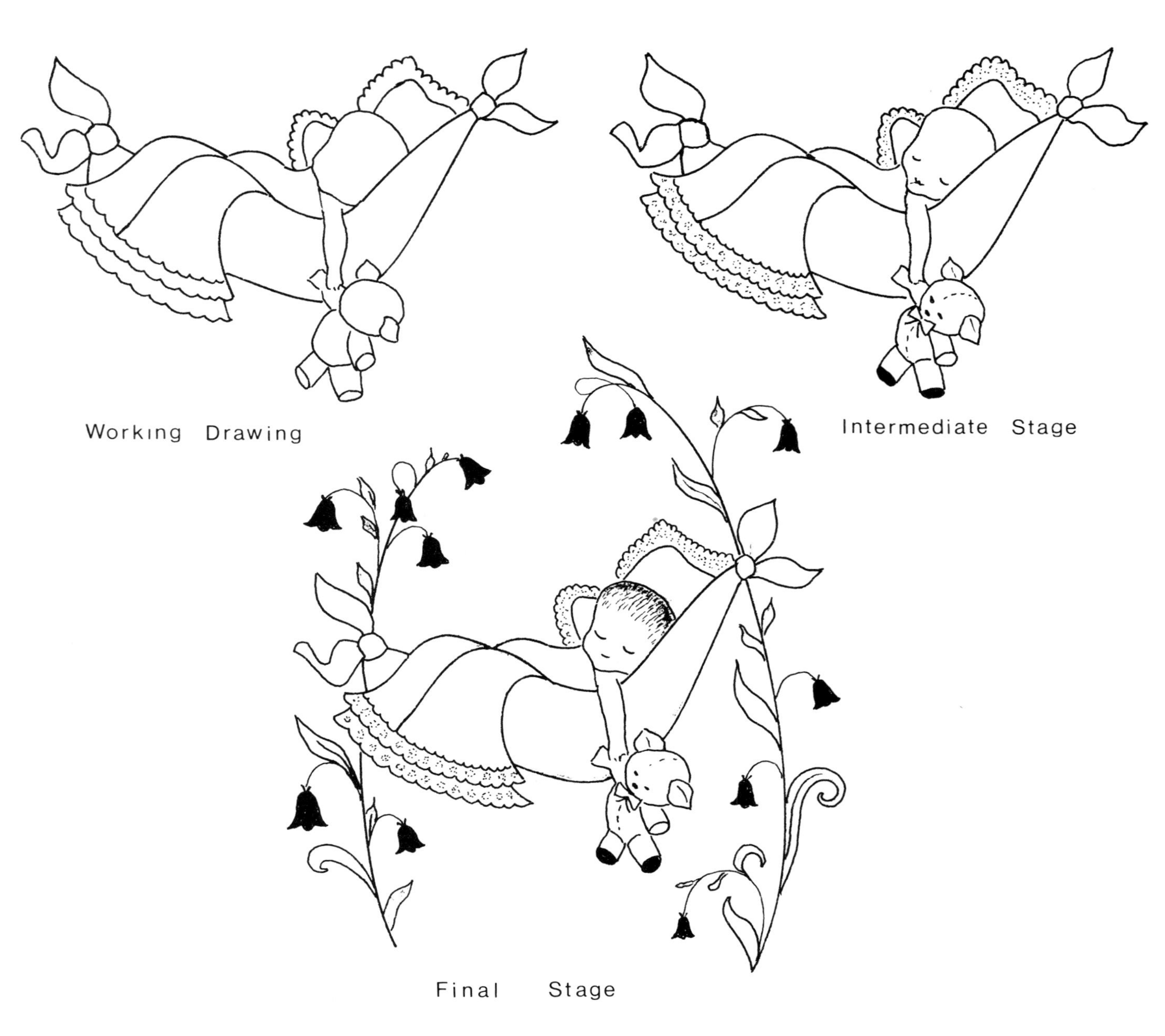

Fig. 86 — *contd.*

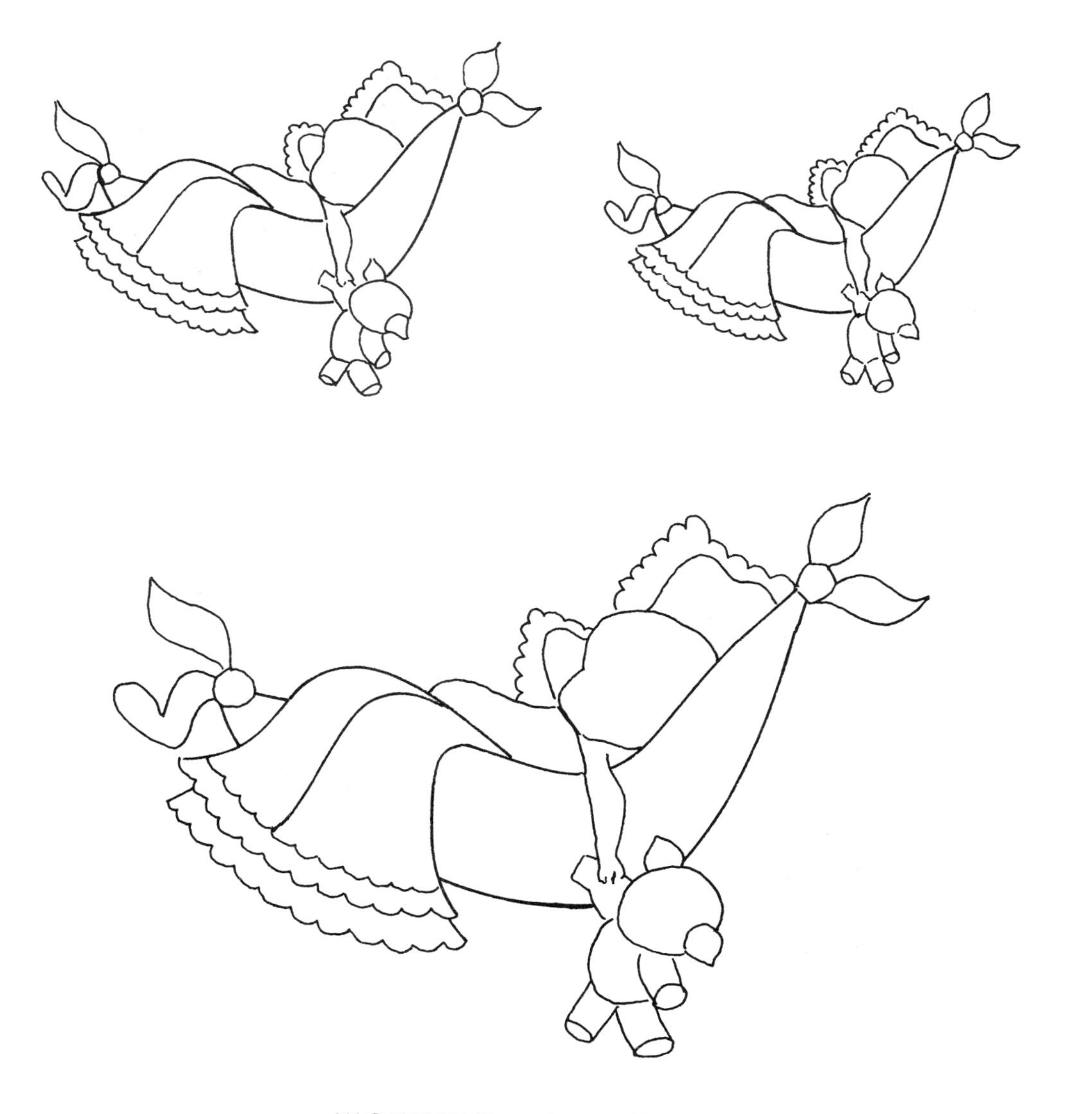

WORKING DRAWINGS

Fig. 86 — *contd.*

Fig. 86 — *contd.* Copyright supplied courtesy of F. J. Warren Ltd.

WORKING DRAWINGS

Fig. 86 — *contd.* Copyright supplied courtesy of F. J. Warren Ltd.

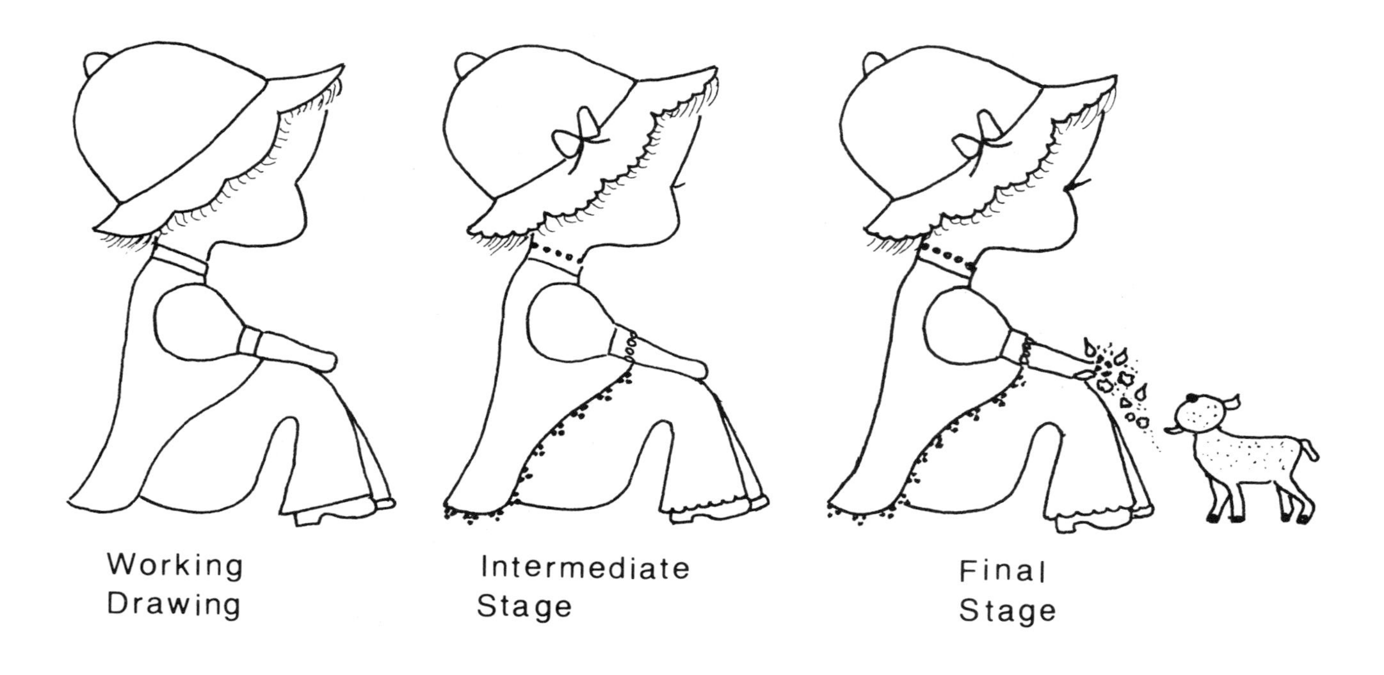

WORKING DRAWINGS

Fig. 86 — *contd.*

Fig. 86 — *contd.*

brown. This should be tried on a separate surface until the desired shade is achieved.

Brush marks in the icing indicate the texture of hair on an all white figure. These can be made when the hair is flooded or by adding a thin layer of softened icing after the runout has dried.

Piping

Additional piping on runouts is always carried out with a No. 0 tube after the runouts are dry.

A little silver paint applied to an edging after the piping has dried is very attractive on a bride's head-dress, veil or bouquet. Very little paint should be used and it must be pointed out that it is not edible.

Flowers in a child's hand can not be piped until after the runout has been placed on the cake.

Procedure for Runout Figure Piping

The procedure is illustrated with step-by-step photographs in Figs 79 to 85 and Plate 7. These illustrations together with the following comments, will enable you to carry out these and other designs.

1. Examine drawings carefully before starting work.
2. Prominent features should be piped first to give the figure a three dimensional form.
3. Leave figures under a lamp for as long as possible to assist the drying and to ensure a pleasing gloss.

Fig. 86 contains many drawings for runout figures that are suitable for tracing from the book.

CHAPTER 12

Different Styles of Decorating

When a style is described as 'going out of fashion' our thoughts turn towards clothes. It is true that styles of dress are constantly changing: to a lesser extent, the same can be said about cake decorating.

In general, the styles of today are plainer than those of the past when piping was ornate and collars (usually more than one) very elaborate.

It is certain that old styles will come back; probably not the same as before, but requiring the same techniques. It is for this reason that two different styles of royal icing are included in this chapter.

The first is a description of trellis work based on the work of Marjorie Berry who is a specialist in this form of decoration.

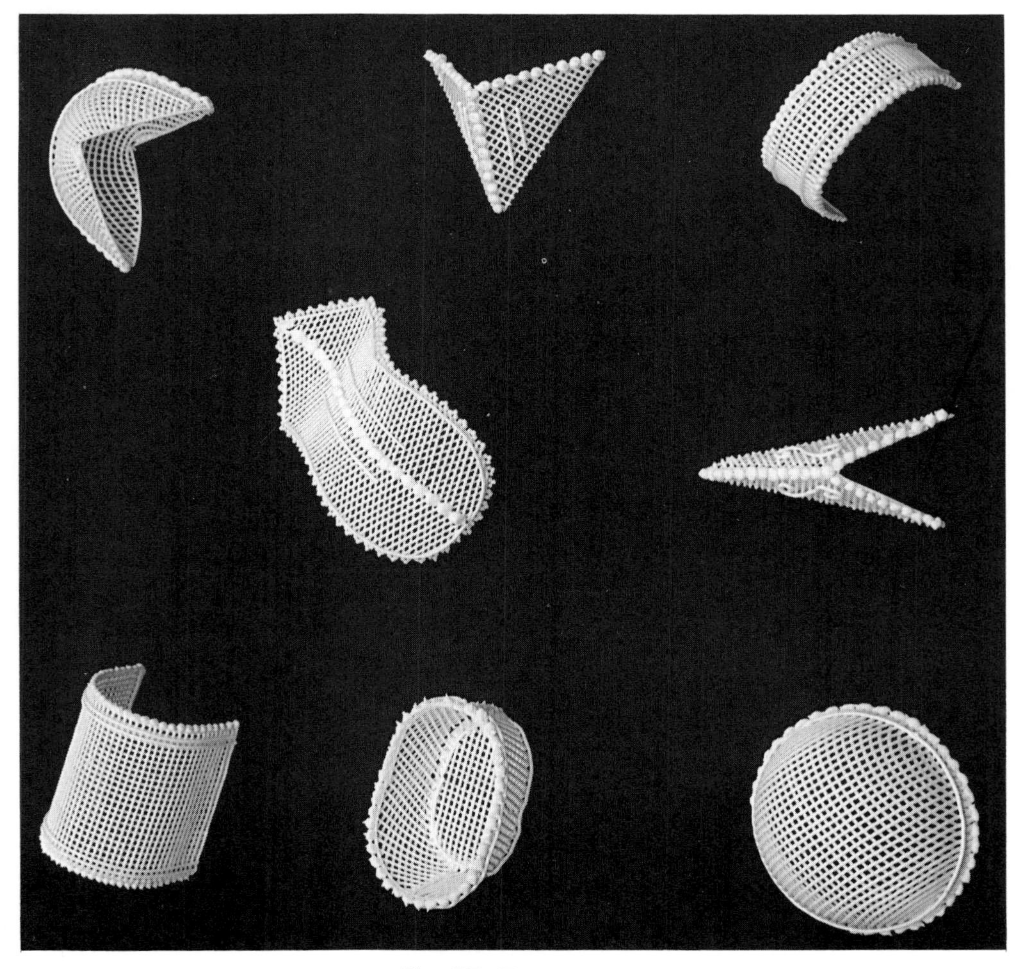

Fig. 87. Trellis work.

Fig. 88. Cake with trellis work — courtesy of M. Berry.

The second illustrates double collars and includes several designs that can be traced from this book.

Trellis Work

This is a decorative form of royal icing that can be carried out with white or coloured icing.

There are four main methods of application.

1. Piped directly onto the cake, the dome appearance being obtained by piping several layers on top of each other.
2. Piped onto net or tulle to enable a decorative motif to be used in the cake design.
3. Extension work incorporated on cakes coated with sugar paste.
4. Piped onto moulds and transferred to the cake when dried. Moulds can be bought to form several of the items shown in Fig. 87. Alternatively, household items such as cream horn tins, jam jars, serviette rings etc. can be used to form the desired shape.

The cake in Fig. 88 is decorated with items that have been piped onto moulds and the method used is outlined here.

(a) Smear the surface of the mould with a *little* white fat. (The exception to this was the 'S' shaped decoration on the corners of the cake (Fig. 88). This mould, made from a piece of metal, was covered with thin waxed paper before the design was piped).
(b) Pipe a line around the base of the mould with a No. 1 tube using the *touch*, *lift* and *place* method. The icing must not contain glycerine.
(c) Pipe a line along the centre of the form.
(d) Pipe lines on either side of this centre line keeping them as close as possible but not touching. Continue piping until the desired pattern has been achieved. Neatness is most important as it affects the overall appearance. Any icing removed from the mould must not be re-used as the grease will affect the mixture.
(e) Leave for twenty-four hours to dry.
(f) When thoroughly dry, the mould is held upright approximately 15 cm (6 in) above a low heat on a gas or electric ring, until the fat melts and releases the icing.
(g) Items made on moulds are attached to the cake with a small amount of royal icing. Joins are covered by trimming with small bulbs of royal icing.

Double Collars

It should be possible to carry out the designs for both collars (round and square) by following the illustrations relating to the square collar (Figs 89 to 92).

Fig. 89. Under collar for square cake.

Fig. 90. Under collar with linework.

Fig. 91. Piping for top collar.

Fig. 92. Completed top collar.

Fig. 93. Piped and flooded numerals.

Fig. 94. Runout bells (5 stages) — stage 1.

Fig. 95. Runout bells (5 stages) — stage 2.

Fig. 96. Runout bells (5 stages) — stage 3.

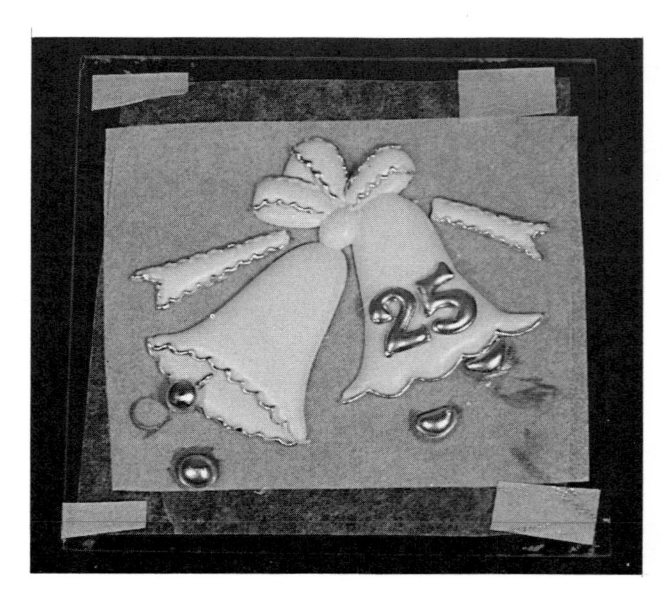

Fig. 97. Runout bells (5 stages) — stage 4.

Fig. 98. Runout bells (5 stages) — stage 5.

Fig. 99. Silver wedding cake with double collar (square).

Square Cake (*Fig. 99*)

This is a 20 cm (8 in.) cake placed on a 30 cm (12 in.) board.

Details of flooding the cake board are shown in Fig. 49.

The marzipan roses on the side of the cake are deep pink and the template is shown in Fig. 41.

Fig. 101. Silver wedding cake with double collar (round).

The design for the corner pieces is shown in Fig. 57.

The *under collar* (Figs 89 and 90) is a deeper shade of pink than the top collar and coating.

(*text continued p. 186*)

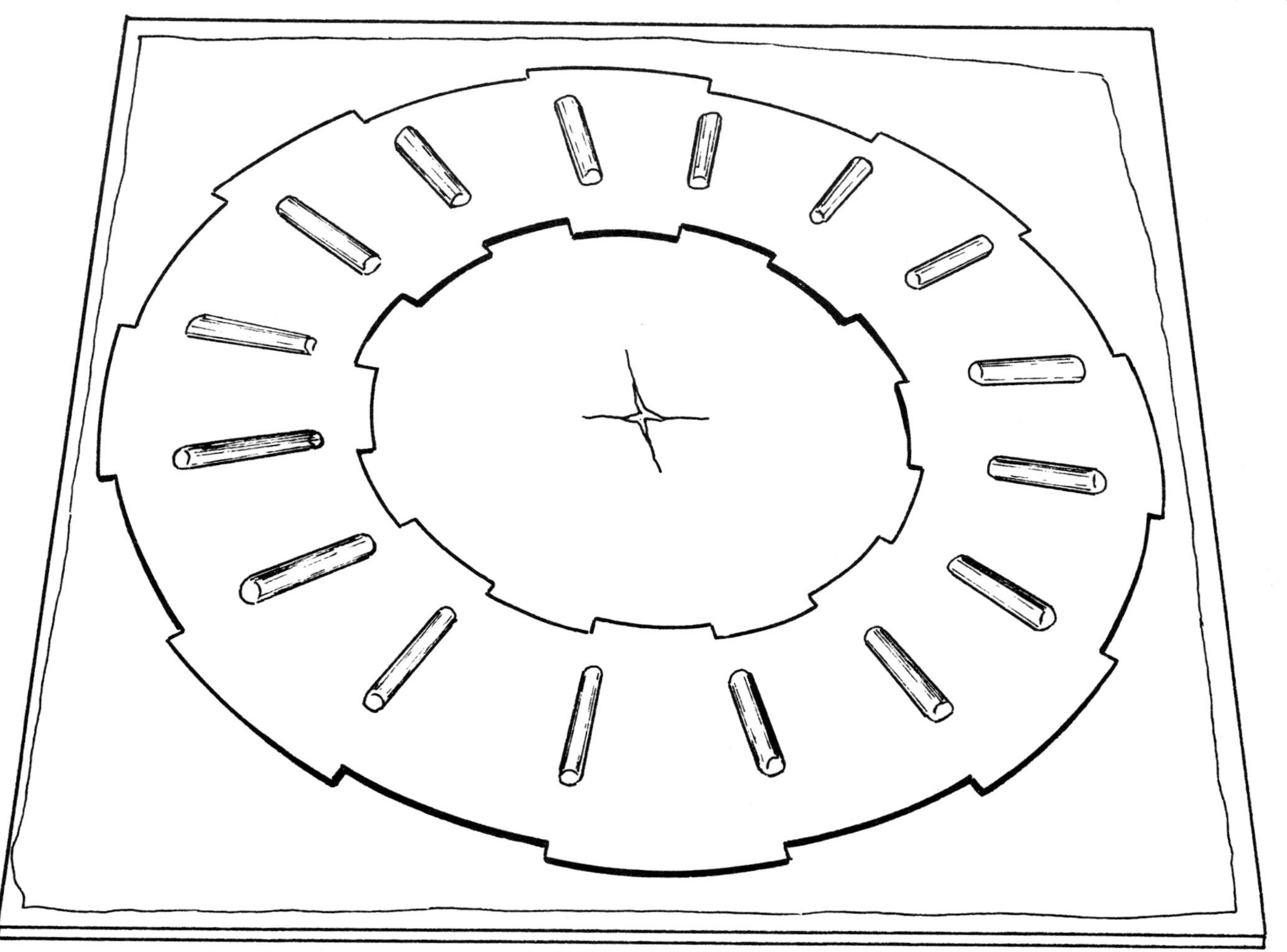

Fig. 100. Under collar with linework (round cake).

Fig. 102. Working drawings for double collars. Above: top collar for 20 cm (8 in.) cake — half of pattern at full size for tracing. See opposite for whole collar pattern.

Fig. 102 — *contd.* Pattern reduced in size to show whole of collar.

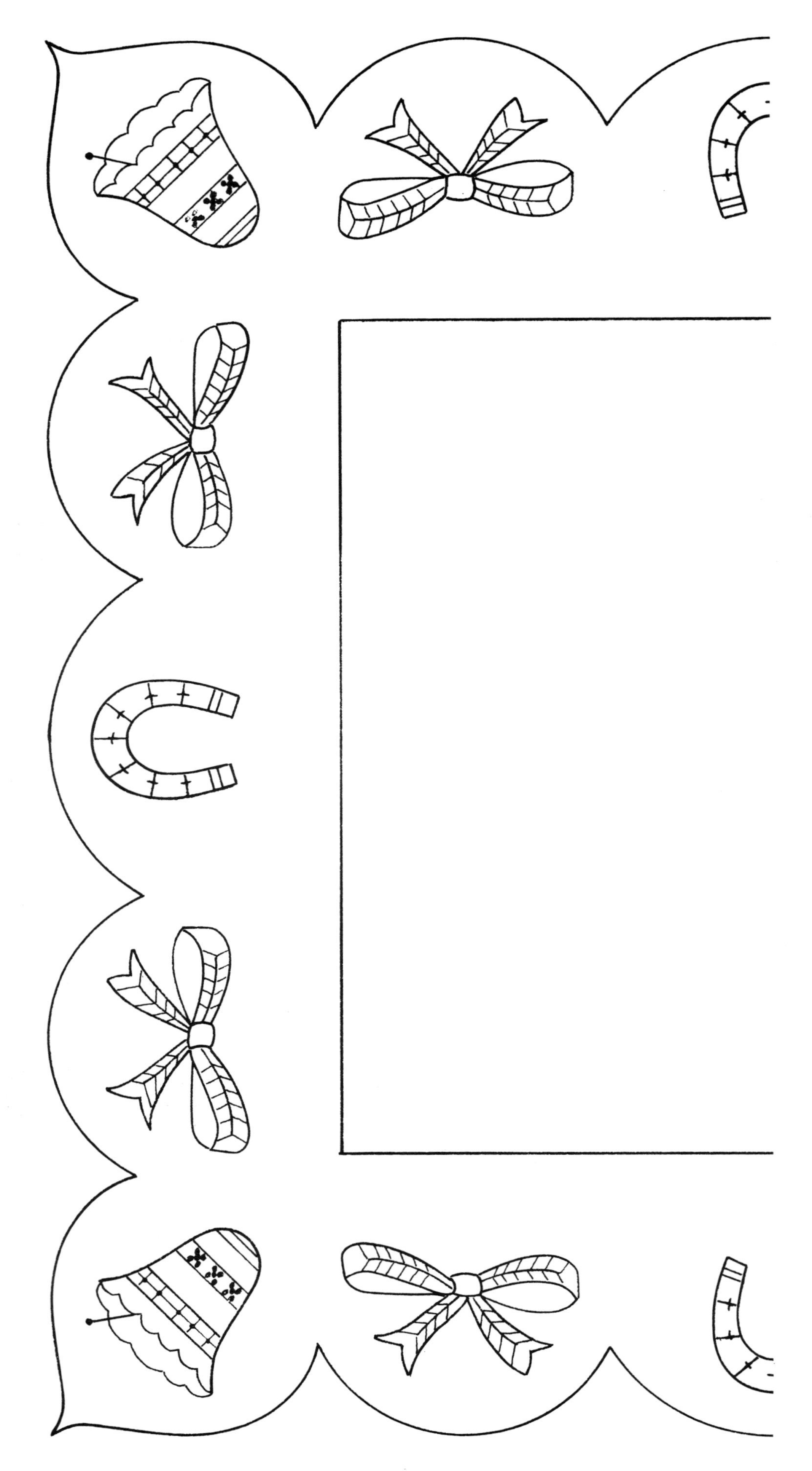

Fig. 102 — *contd.* Top collar for 20 cm (8 in.) cake — half of pattern at full size for tracing. See opposite for whole collar pattern.

Fig. 102 — *contd.* Pattern reduced in size to show whole of collar.

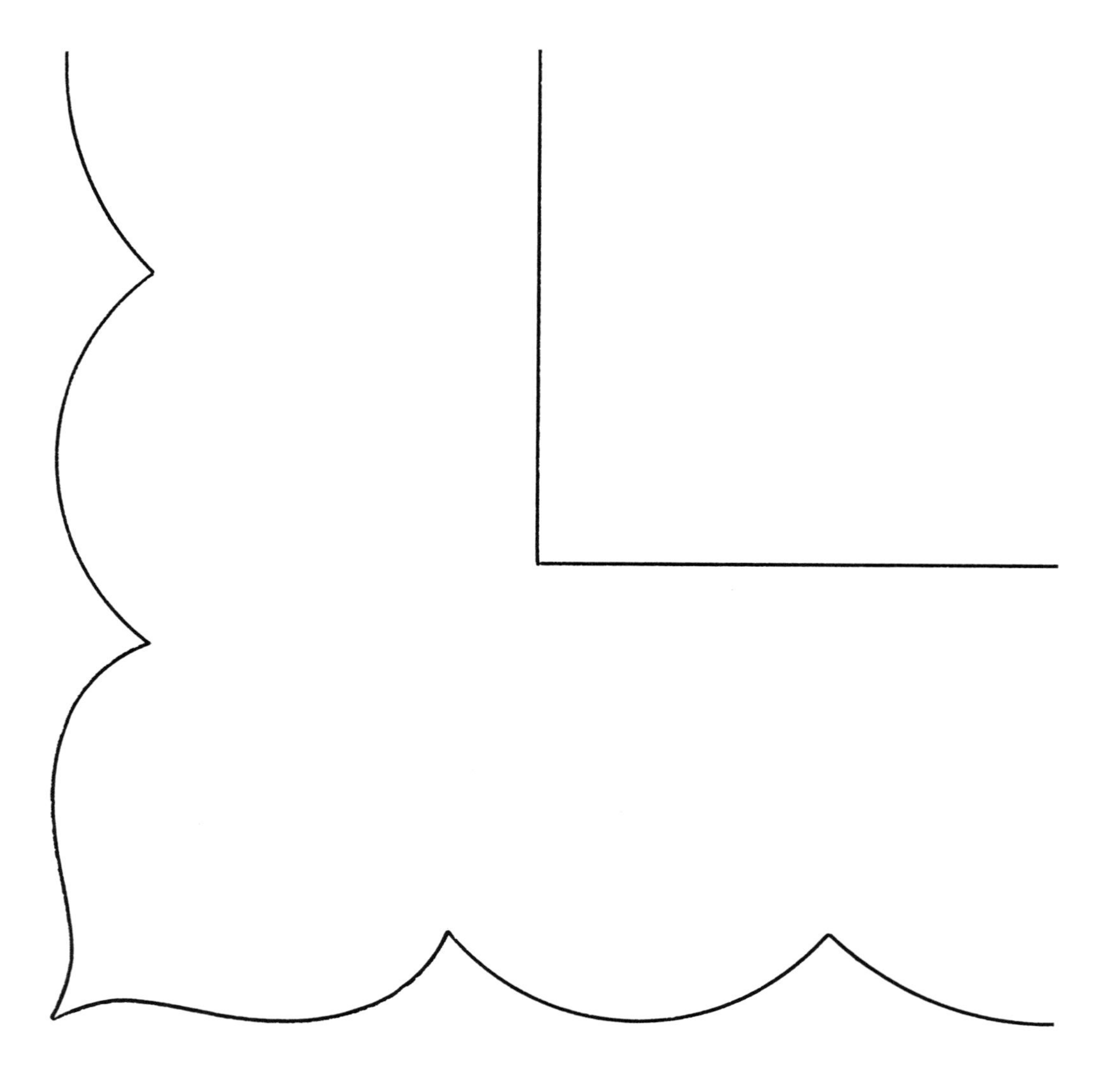

Fig. 102. — *contd.* Under collar for 20 cm (8 in.) cake — quarter of pattern at full size for tracing. See opposite for whole collar pattern.

Fig. 102 — *contd.* Pattern reduced in size to show whole of collar.

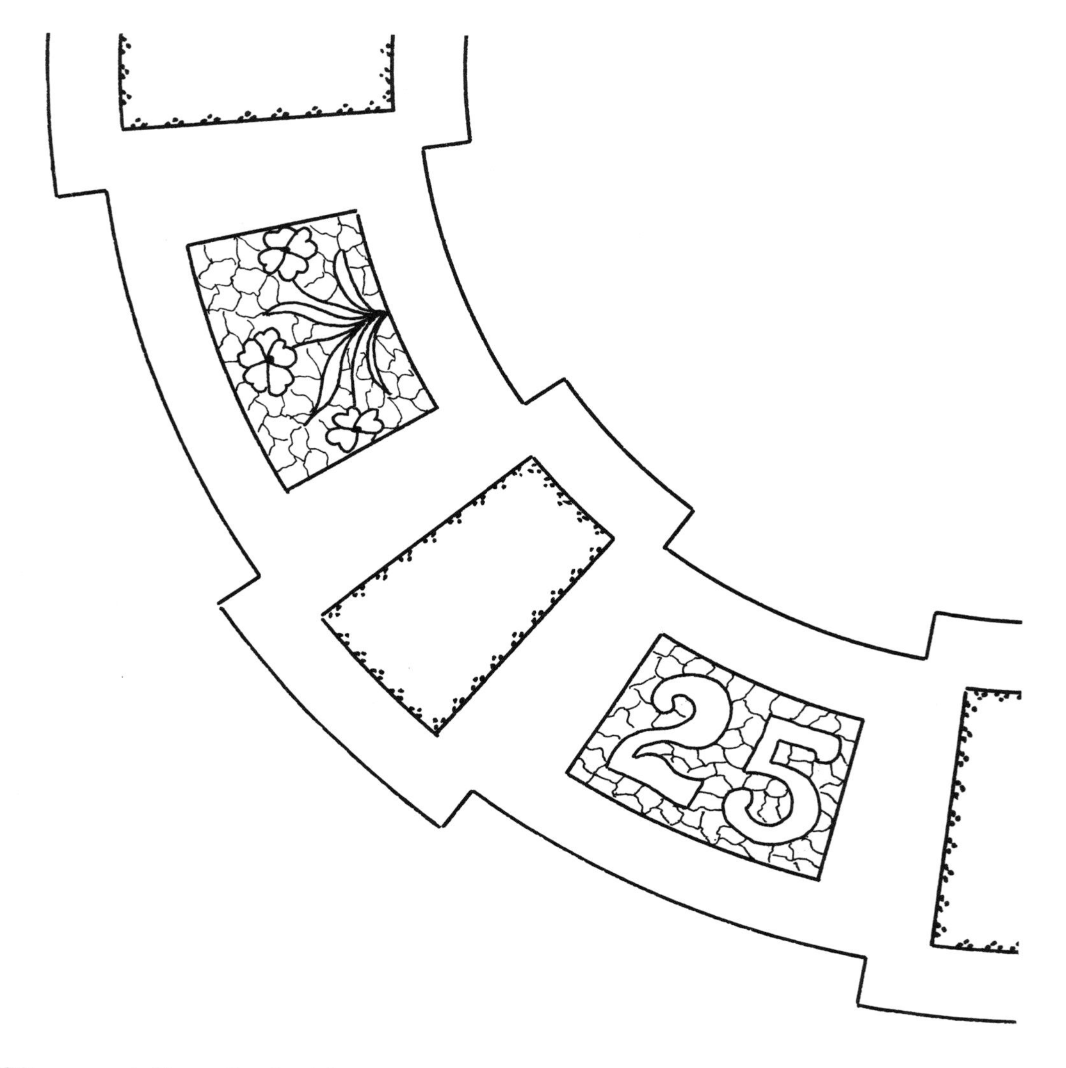

Fig. 102. — *contd.* Top collar for 20 cm (8 in.) cake — quarter of pattern at full size for tracing. See opposite for whole collar pattern.

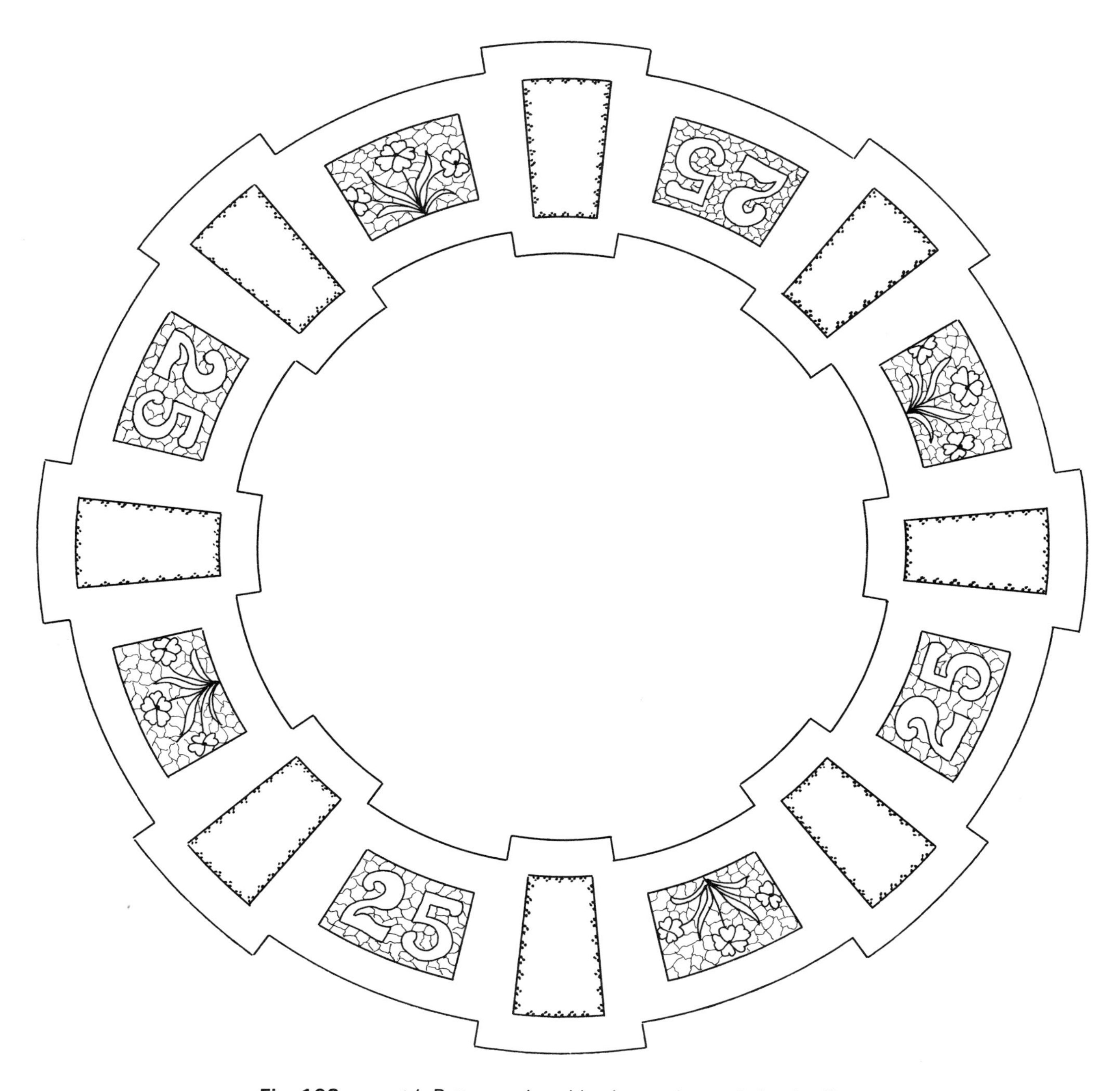

Fig. 102 — *contd.* Pattern reduced in size to show whole of collar.

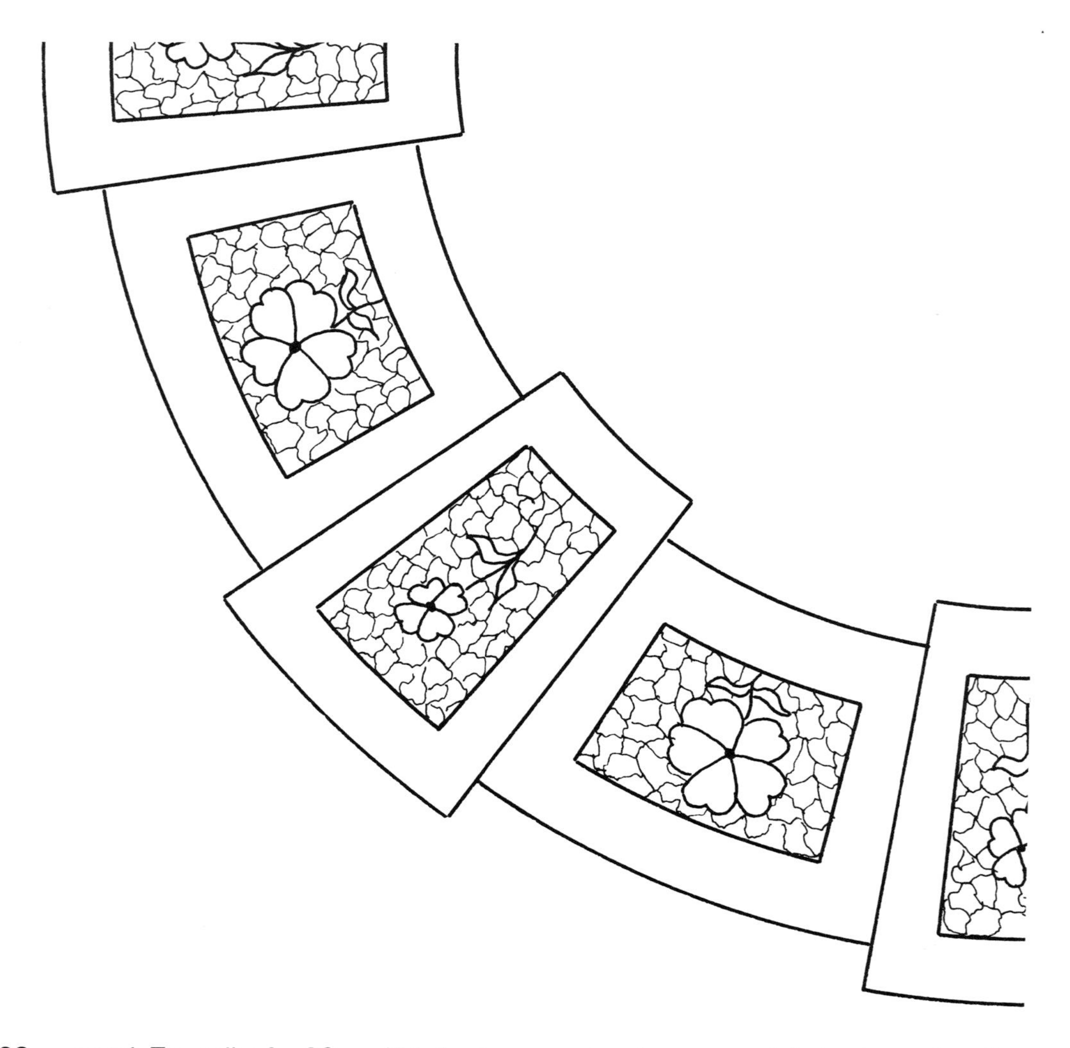

Fig. 102. — *contd.* Top collar for 20 cm (8 in.) cake — quarter of pattern at full size for tracing. See opposite for whole collar pattern.

Fig. 102 — *contd.* Pattern reduced in size to show whole of collar.

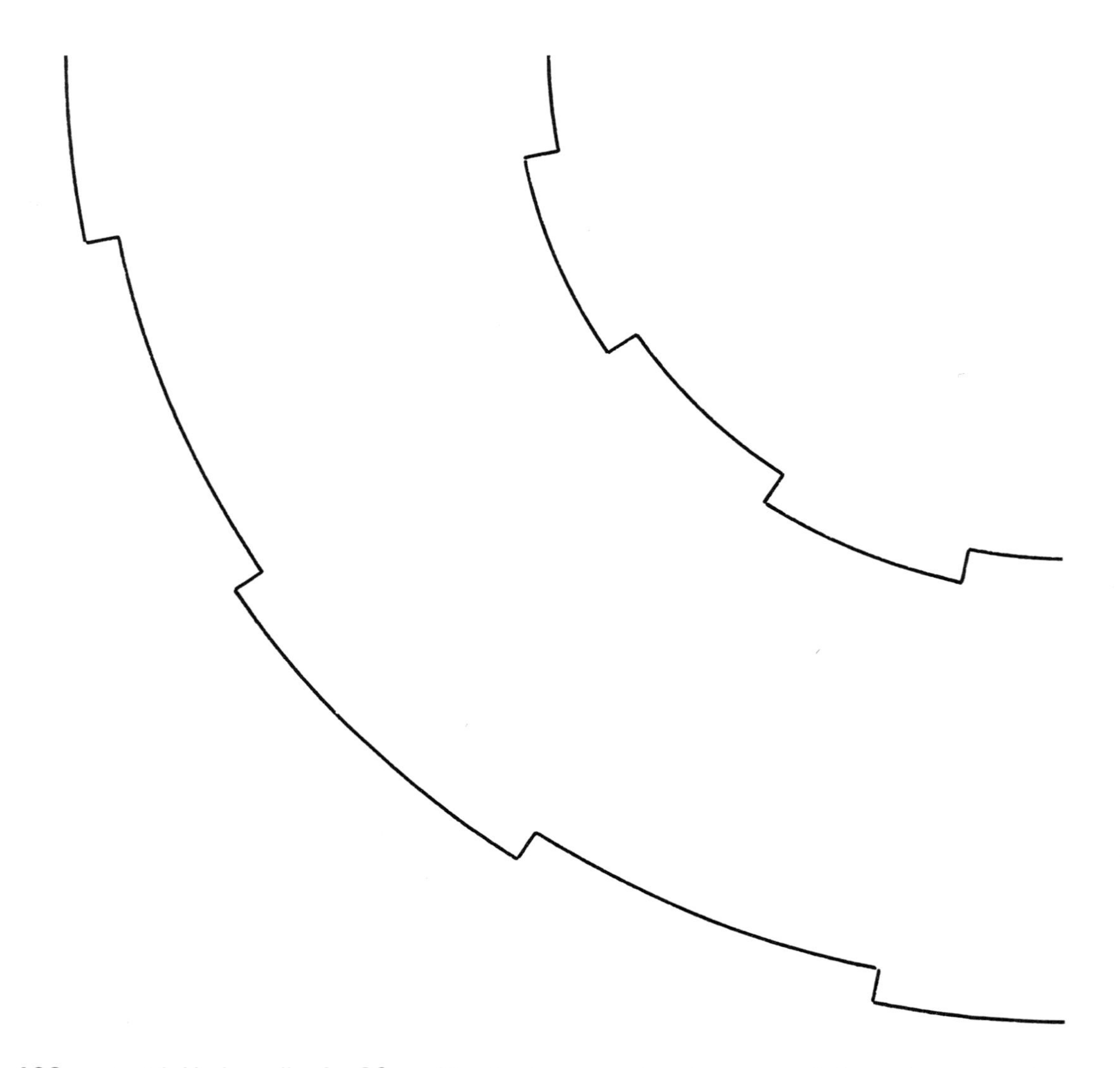

Fig. 102. — *contd.* Under collar for 20 cm (8 in.) cake — quarter of pattern at full size for tracing. See opposite for whole collar pattern.

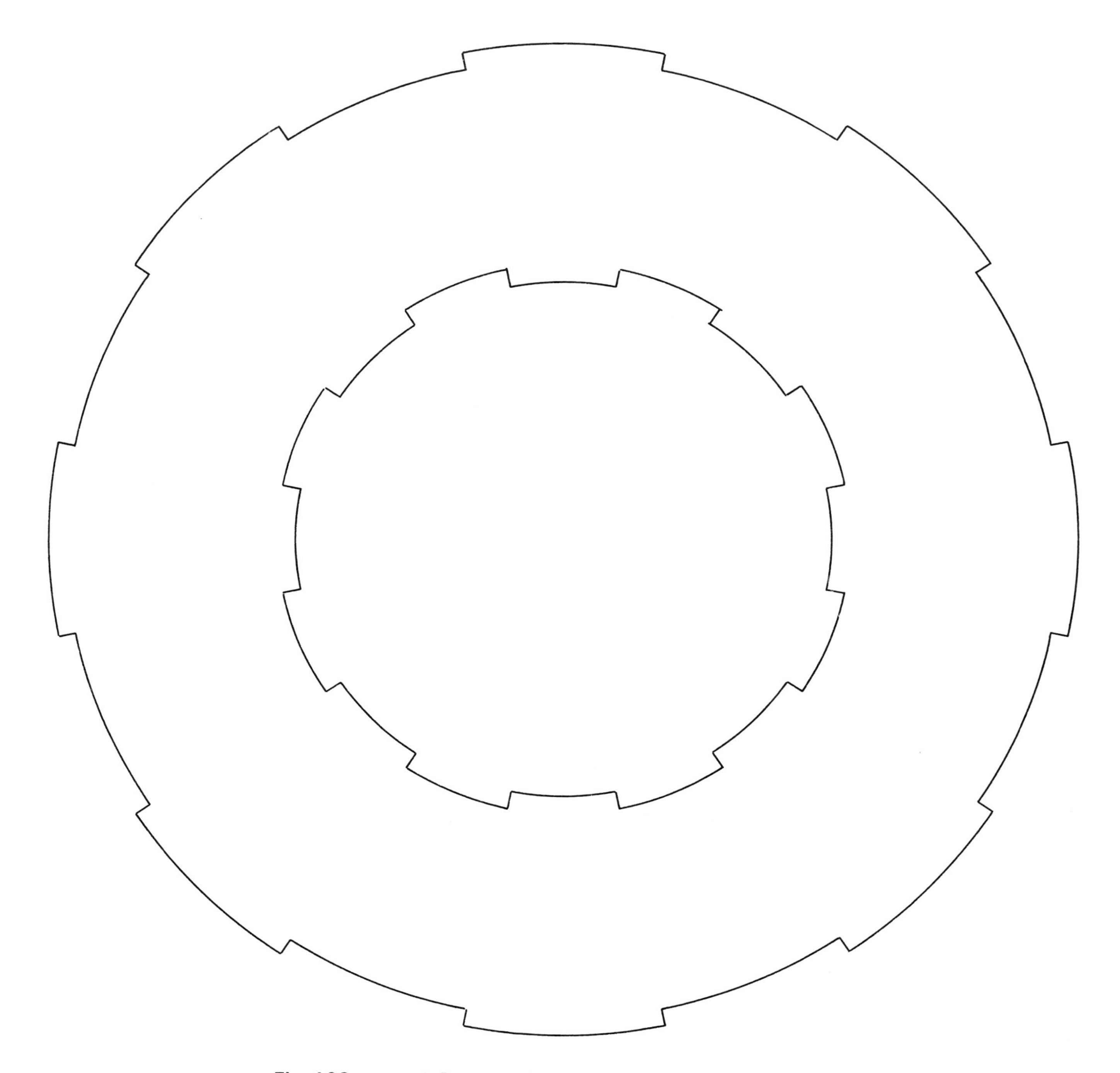

Fig. 102 — *contd.* Pattern reduced in size to show whole of collar.

The *top collar* (Figs 91 and 92) was positioned on the under collar before both were placed on the cake. This was done by piping a line with a No. 3 tube and allowing it to dry (Fig. 90). The line was overpiped with a No. 2 tube and the top collar assembled before the icing was allowed to crust.

Top decoration

Bells, numerals and a plaque are used to decorate this cake and the order of piping is shown in Figs 93–98. Working drawings for the bells are shown in Fig. 86. The completed cake is shown in Fig. 99.

Round Cake (Fig. 101)

This is a 20 cm (8 in.) cake placed on a 30 cm (12 in.) board.

The cake coating and the top collar are white with peach coloured under collar. Small peach coloured flowers and green leaves are contained in the top collar and in the side linework.

The No. 3 linework on the under collar was piped in a different position to that on the square cake and this is illustrated in Fig. 100.

The completed cake is shown in Fig. 101.

Top decoration

A flower is used to decorate the top of this cake. Details of how to pipe this are shown on page 36 and Fig. 51. The working drawings for the petals are given in Fig. 53.

Fig. 102 contains drawings for double collars.

CHAPTER 13

Wedding Cakes

A wedding cake attracts more attention than any other cake and for this reason deserves special attention.

There should be no difficulty in icing a tiered cake, provided the basic principles which apply to a single cake are followed. All too often, however, we hear stories of cakes with rock-hard icing; cakes that lean precariously and, worse still, cakes that actually fall down. This is not only embarrassing but can spoil a special occasion and result in damaging the cake decorator's reputation. Ways of avoiding such mishaps are outlined here.

1. A rich fruit cake should be moist but not wet.
2. Cakes must be level before they are pasted.
3. Glycerine may be added, if required, for icing used for coating.
4. Cake pillars must be evenly spaced.
5. Extra care is required when moving a large cake.

Each tier of a wedding cake must be trimmed level (if necessary) before it is pasted. If the cakes are not right at this stage they never will be. If the cakes are level, the icing does not need to be rock hard because the flat surface, plus even distribution of weight on the cake pillars will ensure against accidents.

There is no need to ice a wedding cake several weeks in advance of the required date. Runout work should be carried out as soon as possible but the decoration need not be completed until about a week prior to the wedding.

Worries about the cake falling down are easily dispelled. When the cake has been coated, assemble the tiers and leave it undisturbed for several hours. In the unlikely event of something being amiss, there will still be time to put it right; it is too late on the day! After dismantling the cake it can be covered with a sheet of fine white tissue paper, if necessary, until it is decorated.

A large cake should be handled with care. Do not lift a cake by taking hold of the edge of the cake board as any flexing would cause the icing to crack. Slide the cake onto the palm of your hand to give maximum support.

The cake is the responsibility of the decorator even if it has been made by someone else. Since it is unlikely that a cake made by another person will have been baked perfectly, be level, be the right diameter, be the correct depth etc.; it is inadvisable to decorate a cake that you have not made. It is your reputation that is at stake.

Decorating a Cake

The order of decoration set out here is a guide applicable to all cakes with, or without tiers.

Round and Square Cakes where a Side Template is Not Required

1. Coat cake and allow to dry.
2. Flood cake board and allow to dry.
3. Place or pipe top decoration.
4. Add decoration, if any, to the top edge of the cake (underneath the border).
5. Edge cake board with No. 0 tube.
6. Assemble top border and pipe inside linework.
7. Place (or pipe) side decoration in line with the top border.
8. Add shell border around the bottom of the cake.
9. Assemble corner pieces (if any) before the shell border has dried. (Where corner pieces

Fig. 103. Two tier round wedding cake.

are to be set into the flooding this takes place immediately after the board has been flooded and before decorating commences.)

Round and Square Cakes where a Side Template is Required

When a template is required for the side of a cake (Fig. 41 and page 22) it is used before the cake board is flooded. Only the essential piping is carried out prior to the template being removed and any further decoration is completed after the board has been flooded and is dry.

Pillars
Cake pillars are available in all shapes and sizes. I prefer to use 9 cm ($3\frac{1}{2}$ in.) pillars on the bottom tier of a cake and 7·5 cm (3 in.) pillars for the middle tier of a three tier cake.

Decorated Wedding Cakes

The designs required to execute the cakes illustrated in this chapter are to be found in this book. Every aspect of piping has been covered and a brief description is given to enable them to be easily reproduced.

Two Tier Round Wedding Cake (*Fig. 103*)

Size of cakes
15 cm (6 in.) round and 25 cm (10 in.) round on 23 cm (9 in.) and 38 cm (15 in.) boards respectively.

Colours, coating
White.

Colours, decoration
White, pink.

Tubes
3, 2, 1 and 0.

Runouts
Border pieces (white) Fig. 68 (8 pieces on each cake). Piped butterflies (white, deep pink) (Fig. 53). Plaque (pale pink with white monogram outlined in silver).

Top decoration
Vase of flowers.

To make a template
Measure the exact length around the cake and cut a piece of paper to suit. Fold the paper into the desired number of sections and cut out a curve.

Divide template for top cake into eight; bottom into sixteen. Essential linework is carried out with a No. 1 tube before the cake boards are flooded.

The plaque holding the monogram is placed on the cake and edged with deep pink bulbs with a No. 0 tube.

Pipe evenly spaced loops with No. 2 and No. 1 tubes at the top edge of the cake.

Edge cake boards with small bulbs of white icing.

Assemble top border pieces in line with the side decoration.

Complete side decoration with piped bells and flowers (No. 1 tube) and alternate deep pink and white butterflies, tipped with silver paint.

Assemble white butterflies on the top borders in line with those on the sides of the cakes.

Pipe a No. 3 shell border at the bottom of the cakes.

Attach cake pillars, evenly spaced, to the cake with a little icing and shell edge with a No. 2 tube.

Glue silver paper banding to the edge of the silver cake boards.

Two Tier Square Wedding Cake (*Fig. 104*)

Size of cakes
15 cm (6 in.) square and 30 cm (12 in) square on 23 cm (9 in.) and 41 cm (16 in.) boards respectively.

Colours, coating
White.

Colours, decoration
White, blue.

Tubes
2, 1 and 0.

Runouts
Border pieces (Fig. 68). Bows (Fig. 61) - outlined with blue. Corner pieces (Fig. 57) with blue piping No. 0 tube. Butterflies (Fig. 61) - two sizes. Piped butterflies (Fig. 53) - white. Plaque - blue with white bride and groom (Fig. 86). Side motifs (Fig. 61) - hearts outlined with blue.

Top decoration
Vase of flowers.

Runout bows were attached to the border pieces with a little icing before the border pieces were removed from the waxed paper.

Flood cake boards using top border piece drawings as a template and assemble corner pieces before the icing is allowed to crust.

Fig. 104. Two tier square wedding cake.

Attach plaque to the cake with a little icing.

Pipe cake boards with filigree using a No. 0 tube and edge with bulbs to match those on the top border pieces.

Assemble top borders and inside linework piped with No. 2 and No. 1 tubes.

Pipe shells underneath the border pieces with a No. 2 tube.

Stick butterfly bodies on the cakes; (in line with the corners of the runout border pieces) assemble wings with a little icing from a No. 1 tube and hold in place with cotton wool until dry.

Place small butterflies between the top borders.

Attach motifs to the sides of the cakes.

No. 2 shell border is piped at the bottom of the cakes.

Evenly space cake pillars and attach to the cake with a little icing and shell edge with a No. 2 tube.

Glue silver paper banding to the edge of the silver cake boards.

Three Tier Round Wedding Cake (Fig. 105)

Size of cakes
13 cm (5 in.), 20 cm (8 in.) and 28 cm (11 in.) round on 20 cm (8 in.), 28 cm (11 in.) and 38 cm (15 in.) round boards respectively.

Colours, coating
White.

Colours, decoration
White, peach.

Tubes
2, 1 and 0.

Runouts
Border pieces – white with peach coloured flowers (Fig. 68); (6 pieces on 13 cm (5 in.), 8 pieces on 20 cm (8 in.) and 10 pieces on 28 cm (11 in.) cake).

Piped butterflies – white and deep peach (Fig. 53). Bells – white edged with silver (Fig. 86). Hearts – white outlined with silver (Fig. 61). Plaques (two sizes) – pale peach with white monograms and piping.

Top decoration
Cluster of peach coloured marzipan roses assembled on a plaque (Fig. 39).

Divide templates to suit the respective tiers into six, eight and ten. Essential linework is carried out with a No. 1 tube before the cake boards are flooded.

Place plaques in the centre of cakes and edge with white bulbs using a No. 0 tube.

Pipe evenly spaced loops with No. 2 and No. 1 tubes at the top edge of the cakes.

Assemble top border pieces in line with the side decoration.

Decorate the sides with piped bells and flowers using No. 1 tubes, and runout hearts and bells. Alternate deep peach and white butterflies complete the decoration.

A No. 2 shell border is piped at the bottom of the cake.

Evenly spaced cake pillars are attached to the cake with a little icing and shell edged with a No. 2 tube.

Glue silver paper banding to the edge of the silver cake boards.

(The cluster of roses is placed on top of the cake after the cake has been assembled. It could have been stuck to the cake and edged with a No. 0 tube at the same time as the plaques.)

Three Tier Square Wedding Cake (Fig. 106)

Size of cakes
13 cm (5 in.), 20 cm (8 in.) and 28 cm (11 in.) square on 20 cm (8 in.), 30 cm (12 in.) and 41 cm (16 in.) boards respectively.

Colours, coating
White.

Colours, decoration
White, deep pink, green.

Tubes
2, 1 and 0.

Runouts
Border pieces – white with pink flowers and green leaves (Fig. 68). Flooded butterflies – white outlined with deep pink (3 sizes) Fig. 61. Piped butterflies – white (Fig. 53). Bows – white oulined with deep pink (3 sizes) (Fig. 61). Corner pieces – white (Fig. 57). Plaques – white with white piping and white monograms outlined with deep pink (2 sizes) (Figs 47, 73 and 75).

Top decoration
Vase of flowers.

Cake boards are flooded using top border piece drawings as templates.

Place plaques in position and edge with No. 0 tube.

Pipe evenly spaced loops using No. 2 and No. 1 tubes at the top edge of cakes.

Edge cake boards with single No. 0 bulbs to match the edging on the top borders.

Fig. 105. Three tier round wedding cake.

Fig. 106. Three tier square wedding cake.

Top borders are assembled and inside linework piped with No. 2 and No. 1 tubes.

Butterfly bodies are stuck on the cakes; (in line with the corners of the runout border pieces) assemble wings with a little icing from a No. 1 tube and hold with cotton wool until dry.

Small butterflies are placed between the top borders.

Bows are attached to the sides of the cakes.

Pipe a No. 2 shell down the corners and around the bottom of the cakes and place the corner pieces in position.

Evenly spaced cake pillars are attached to the cake with a little icing and shell edged with a No. 2 tube.

Glue silver banding to the edge of the silver cake board.

Index